Lecture Notes in Artificial Intelligence 517

Subseries of Lecture Notes in Computer Science
Edited by J. Siekmann

Lecture Notes in Computer Science
Edited by G. Goos and J. Hartmanis

K. Nökel

Temporally Distributed Symptoms in Technical Diagnosis

Springer-Verlag
Berlin Heidelberg New York
London Paris Tokyo
Hong Kong Barcelona
Budapest

Series Editor

Jörg Siekmann
Institut für Informatik, Universität Kaiserslautern
Postfach 3049, W-6750 Kaiserslautern, FRG

Author

Klaus Nökel
Zunftstraße 38a, W-8013 Haar, FRG

CR Subject Classification (1991): I.2.1, I.2.4, I.5.2

ISBN 3-540-54316-3 Springer-Verlag Berlin Heidelberg New York
ISBN 0-387-54316-3 Springer-Verlag New York Berlin Heidelberg

Printing and binding: Druckhaus Beltz, Hemsbach/Bergstr.
2145/3140-543210 - Printed on acid-free paper

Preface

Existing diagnostic expert systems usually operate under a set of simplifying working assumptions which limit their universal applicability. A common assumption concerns the treatment of time-dependent information: the device to be diagnosed is assumed to have a static behavior, i.e. the relation between inputs and outputs is constant over time. In most realistic application domains this assumption is violated and both the normal, intended function of the device and the potential malfunctions are complex behaviors over time. This thesis addresses the problem of systematically treating information about fault symptoms which are spread out over periods of time.

These symptoms are characterized by a specific order of events, and in the general case a single snapshot of the device state does not contain sufficient information to recognize an occurrence of the symptom. Instead one has to plan a measurement sequence that consists of several observations at more than one time point.

Starting with a classification of the various types of dynamic faulty behavior we identify one class (temporally distributed symptoms [TDSs]) for which current expert system technology is particularly deficient. For this class we design a representation language that allows TDSs to be specified in a declarative manner. Based on this representation we discuss the problem of giving a meaningful definition of a "successful match of a measurement sequence against a TDS specification". We then operationalize our definition in the form of an algorithm which plans such an observation sequence based on the TDS specification.

Finally, we demonstrate that our solution is a generic, paradigm-independent building block for diagnostic expert systems by embedding it into the frameworks of both an associative/heuristic and a model-based diagnostic system.

This book is based on my doctoral dissertation, accepted in fulfilment of the requirements for the degree of Dr. rer. nat. at the University of Kaiserslautern.

I am indebted to many people who helped greatly in the preparation of the thesis and the research leading up to it. First and foremost, I thank my advisor, Prof. Dr. M.M. Richter, for always asking the right questions and thus drawing my attention to new and interesting aspects of my work that I would otherwise have overlooked. Second, a huge thanks to my wife, Scarlet, who unfailingly kept me going even when this meant yet another spoiled evening or weekend. Her continuous motivation was a critical prerequisite for my work. Third, this thesis is the end-product of a complex evolution that benefited greatly from uncountable discussions with my co-advisor, Prof. Dr. W. Dilger, and many colleagues. Among them Peter van Beek, Rainer Decker, Oskar Dressler, Hartmut Freitag, Thomas Guckenbiehl, Henry Kautz, Frank Maurer, Robert Rehbold, Gisela Schäfer-Richter, Peter Struss, and Hans Voss deserve special mention. Fourth, Klaus Becker, Sabine Kockskämper, Hans Lamberti, Johannes Stein, and Stefan Wess carried out much of the actual implementation work and helped me in the debugging of my conceptual design by putting their fingers on all the vague spots. Finally, all the people in the expert systems group at the University of Kaiserslautern provided an extraordinarily pleasant working environment.

Kaiserslautern Klaus Nökel
March 1991

Table of Contents

5 Conclusion

Appendix

1 Introduction

1.1 Technical Diagnosis

Since the dawn of expert systems diagnosis has been one of their predominant tasks. The study of diagnostic procedures and of various kinds of diagnostic knowledge has spawned the development of many techniques (e.g. associative rules, theories for uncertain or vague knowledge) that are now considered an integral part of AI technology and have been applied successfully in many other fields of AI.

Although the first systems dealt with the diagnosis of human diseases, it was soon discovered that the same techniques could also be applied to find faults in all kinds of machines. Systems such as EL [Stallman/Sussman79], MYCIN [Shortliffe76] and ABEL [Patil81] mark the beginning of a very fruitful development that has led to innumerable diagnostic experts systems. Several special journals and conferences mirror both the scientific importance of technical diagnosis for AI and its great commercial impact.

Despite its similarity to medical diagnosis technical diagnosis differs from the former in many essential aspects which are all rooted in the fundamental dichotomy of natural systems vs. artifacts. Since artifacts have been designed by engineers, we have — at least in principle — complete knowledge about their structural composition and functionality. In biological systems this knowledge is largely empiric in nature and only in rare cases it is backed up by a thorough understanding of the connection between structure and function. The choice of tests is also based on different criteria: although in both cases the information gain is of prime importance, medical considerations such as side effects or causing pain to the patient further reduce the range of potential tests whereas in technical systems the economic utility of performing a test (quality of diagnosis vs. test cost and cost of additional standstill time) has to be taken into account. Similarly, the quality of a diagnosis depends directly on the possible treatments. In the medical case a diagnosis is adequate if it is sufficiently exact to indicate a treatment and to minimize the danger of an incorrect diagnosis. The first condition has several subtle implications, e.g. the diagnosis must be found *in time* to start the treatment. In the technical domain incorrect diagnoses or diagnoses which are not reached within a certain time are usually not fatal[1], but expensive. With this in mind many machines are constructed from individually replaceable modules; as soon as the fault is pinpointed within one module the whole module is replaced and the diagnosis can continue while the machine is working again.

In the past years many of these characteristics have been addressed in a systematic way. Technical diagnosis has matured in the sense that there are a number of accepted frameworks for diagnosis (e.g. GDE [de Kleer/Williams] or associative diagnosis (cf. [Richter89])) and it is common to clarify and evaluate new results with respect to these frameworks. Still, these frameworks are evolving: In order to cope with the immense complexity of the task all diagnostic system projects – among them the systems from which the standard frameworks emerged – chose to concentrate only on some aspects. By judiciously selecting application domains in which not *all* "nasty" facets of the diagnostic process play a role, they deliberately simplified the task. This was probably the only way to obtain any positive results, but one must always be aware that these results depend on a number of working assumptions. Consequently, ongoing research is aimed in part at studying the effects of dropping one or more of these assumptions and, if the effects are not satisfactory, of augmenting the basic machinery to solve or at least sidestep the problem.

1.2 Static vs. Dynamic Systems

One of the most popular assumptions is to restrict oneself to static systems. As the use of the word "static" here is somewhat different from everyday use, a few explanations are in order. As [Leitch/Wiegand89] point out, in virtually all physical systems inputs, outputs and internal variables are functions of time. Since no energy transfer happens truly instantaneously, output values are not only affected by the current input values but also by the history of the system. However, in many interesting applications the temporal extent of this memory effect is short in comparison to the time scale on which the system's behavior evolves. For these systems one can find useful approximations which abstract from time and model the relations between inputs and outputs as unchanging equations. They are therefore called static systems, even though they

---

[1] with some important exceptions, such as monitoring of power plants and similar tasks.

possess a behavior that varies over time. Pure combinatoric electronic circuits are well-known examples of this type of system and have – for this very property – been the favorite domain in model-based diagnosis.

Dynamic systems, on the other hand, make up the majority of machines and artifacts in our world. Their state can change, even if the input is kept constant; conversely, the output depends not only on the present but also on past values of the inputs. Not surprisingly, models of dynamic systems are usually more complex than those of static systems. Very often they consist of a system of differential equations or a qualitative approximation.

What is true of dynamic systems in general holds also for their faults. In general fault symptoms in dynamic systems are deviations from the normal behavior which develop *over time*. They fall into three categories:

- *Small-scale phenomena:* these are the effects which happen on a time scale on which it is not feasible to determine the form of the signal by individual measurements. These phenomena (e.g. vibrations) are usually treated in the form of their temporal abstractions (e.g. frequency).
- *Medium-scale phenomena:* the time scale of the effects in this class is on an order that it makes it feasible to detect their occurrence using individual observations.
- *Large-scale phenomena:* some effects such as tool wear evolve over intervals which are considerably longer than the duration of a diagnostic session. These are the effects where statistical methods such as time series analysis [Box/Jenkins70] come into play as a means of preventive action.

1.3 Temporally Distributed Symptoms (TDSs)

In this thesis we shall deal with medium-scale phenomena exclusively. We will refer to these effects as *temporally distributed symptoms* (TDSs) throughout the text. In addition we consider only symptoms which can be given a precise definition, unlike certain TDSs in highly complex chaotic application domains such as weather prediction.

Static systems hold a special appeal for diagnosis because all three types of dynamic misbehavior can simply be neglected: once a fault has occurred the system can be kept and studied in the state which manifests the fault for an indefinite period. Thus the diagnostic system needs to determine only *which* observations should be made and is relieved from reasoning about *when* they should be made. In a static system any discrepancy between inputs and outputs is guaranteed to have an explanation *in the same state*, hence all measurements take place quasi-simultaneously in this state and their temporal order plays no role. In contrast a dynamic system might well look inconspicuous – except for the discrepancy itself –, because the malfunction originated in the past and may still be developing. The techniques one has to employ in order to detect these kinds of faults in a dynamic system depend on the time scale of the symptoms. For TDSs one must perform an experiment based on the current fault hypothesis that involves

- setting up a particular initial state of the system by selecting input values and/or modifying its structure (and, hence, its functionality),
- guiding the system through some sequence of states by varying the inputs according to specific histories,
- predicting behaviors for some set of outputs which for this state sequence are different under the fault hypothesis and normality assumptions, respectively,
- planning observations of these outputs in such a way that exactly the significant deviations are detected.

Obviously, designing such an experiment is in itself a much harder task than figuring out the best candidate from a set of potential measurement positions. In the first two steps it may not even be possible to find a suitable initial state and a state sequence from there to an observable discrepancy, because they would require a physically impossible combination of inputs and actions[2], whereas in "static diagnosis" the state already exists. Predicting the output values in turn requires simulating the model of the system which is much more complicated in the dynamic case. Lastly, planning the observation is trivial in static diagnosis once one has decided on the place of the measurement, because the predicted "behavior" of the measured quantity consists of a single value for each hypothesis. Again, in the dynamic context the predictions are really behaviors over time for which matching has to be defined in the first place and for which special algorithms must be devised.

Indeed the added complexities in dealing with faults in dynamic systems seem quite compelling to try to get by with static approximations of the true system. Diagnostic

---

[2] conflicting in time, of course, i.e. two requirements clash only if they prescribe different values for the same variable at the same time point.

systems operating in this philosophy not only make use of the broad consensus that stepwise discrimination between diagnoses can be achieved by proposing a series of individual measurements and interpreting their results; most of them have gone further than that and have made two additional assumptions:

- Each measurement result contributes *directly* to the discrimination process.
- Based on the measurements taken so far one can always propose a *single* successor measurement[3].

Although these assumptions are justified in the case of truly static systems, we will see in a moment that they lead to awkward diagnostic behavior, if individual measurements are misused to mimic dynamic experiments. When we designed MOLTKE [Althoff et al. 88], an associative expert system that aids in the fault diagnosis of CNC machining centers, we started out with a static view of the machine state. Later, during knowledge acquisition we came across symptoms[4] which could not be described adequately using the representation language we had developed. Finding a solution to this dilemma spawned our work on TDSs; we will therefore draw the generic examples for our techniques from MOLTKE's domain.

Soon we discovered more examples in everyday domains. Consider e.g. that you need to test whether one of the cylinders in your automobile's engine is not working properly. The usual procedure applied to each cylinder in turn would be to observe the speed of rotation with the motor running idle, and then to remove the lead from the sparking plug. A subsequent drop in in the speed of rotation would indicate that the cylinder had been working alright, whereas a constant speed would imply that the cylinder had not been working all along.

In this example, three aspects are worth noting:

- The temporal order of the two measurements and the action is highly significant. This is typical of measurements which in isolation carry no information *except* through the interpretation of other measurements in their context.

---

[3] The claim is not that there will be one unambiguous proposal but that the planning horizon is confined to exactly *one* more measurement.

[4] We will return to one of these original examples in section 3.1.

- As a consequence it seems awkward to suggest the three steps of the test one at a time. Neither of the measurements, much less the action, seem promising, if one does not have the overall effect in mind.

- The action modifies the device under consideration (and hence the model). This means that even if we were willing to stretch the meaning of "measurement" to encompass complex sequences of actions and observations, it would seem unclear how algorithms designed to find "the best location for the next measurement" could handle distributed measurements, since these cannot be assigned a meaningful "location" w.r.t. any one model.

Which conclusions can we draw from these observations regarding the necessity of dynamic experiments? The answer depends on the kind of devices we are trying to diagnose. Particularly where the locations of potential measurements lie relatively dense, measurements are cheap, and the system changes state very slowly or not at all[5], isolated observations are often sufficient. At the other extreme, there are cases where potentially useful measurements are in practice impossible (e.g. because they are destructive) and an observation over time may be the only way to deduce the results from more accessible sources. In between there is a wide spectrum of examples where temporally distributed symptoms are used as substitutes for isolated observations, not because the latter are infeasible in the strict sense, but simply because the former are more convenient (less costly, easier to check, ...). The car engine example belongs in this category. A diagnostic system that aims to choose a series of observations minimizing some measure of cost should reproduce this behavior of human experts.

1.4 Contributions of the Thesis

The goal of this thesis is

- to analyze the various types of faulty dynamic behavior,

- to present a representation language for TDSs,

- to define "matching of temporally distributed observations against the specification of a TDS" in a both principled and useful way,

---

[5] Alternatively, it would be sufficient if states could be reproduced arbitrarily often at no great cost.

- to develop an algorithm that effectively implements this definition, and
- to demonstrate how all these techniques can be integrated within the framework of both an associative/heuristic and a model-based diagnostic system.

1.5 Organization of the Thesis

Following the introduction we survey in chapter 2 how AI systems for diagnostic and non-diagnostic tasks have previously dealt with the representation of time-dependent information and temporal reasoning. At the end of the chapter we will summarize the principal characteristics of the systems discussed and compare them to the approach taken in this thesis.

Chapter 3 contains most of the technical results and falls into seven parts. We start with an informal statement of the problem and give a functional specification of temporal matching in terms of its input/output relation. Next we develop an internal representation for TDSs which supports the operations performed during the matching algorithms. As a side-effect we present a novel characterization of a subset of Allen's interval relations that is sufficiently expressive for our purposes and at the same time possesses attractive complexity properties. The next section deals with the problem of defining temporal matching in such a way that the definition has certain desirable logical properties *and* a realistic implementation. This implementation – the temporal matching algorithm – is described in detail in the next section. At the end of the chapter we propose an extension of our basic approach that allows for the use of certain quantitative temporal constraints in TDS specifications.

In chapter 4 we show how temporal matching can be used as a basic technique in diagnostic expert systems. We demonstrate its generality by integrating it into both the associative/heuristic and the model-based framework. In each case we study an existing system that is typical of the paradigm, MOLTKE and GDE/SHERLOCK, to make the presentation concrete.

Finally, chapter 5 contains a statement about the status of the implementation, an evaluation of our results and a discussion of directions for further research based on our work.

Most of the terminology we use is introduced through formal definitions. There are, however, some concepts which belong to the standard terminology in the literature about temporal reasoning and diagnostic expert systems. Instead of repeating their definitions we merely state which of the established meanings we have in mind; an index at the back of the thesis contains pointers to the first occurrences of these concepts in the text.

1.6 Previous Publications

Some of the material presented in this thesis has been published in the form of conference papers [Nökel89a], [Nökel89b] or has been submitted to journals [Nökel/Lamberti90]. This was thought necessary to provide proof of the originality, because the field has attracted a number of other researchers since the research was started and continues to do so. Indeed some of my results on convex interval relations (section 3.2.5) have been reproduced independently by Peter van Beek [van Beek89] after the publication of [Nökel89a].

Wherever material in this thesis overlaps with previous papers it supersedes the earlier publication. Major efforts have been undertaken to improve both the form of the presentation and the contents; in many cases the terminology has been changed to avoid name clashes and some definitions have been altered in order to improve the readability of the theorems and proofs.

2 Related Work

In 1982 Drew McDermott wrote in his classic paper:

> "A common disclaimer by an AI author is that he has neglected temporal considerations to avoid complication. The implication is nearly made that adding a temporal dimension to the research (on engineering, medical diagnosis, etc.) would be a familiar but tedious exercise that would obscure the new material presented by the author. Actually, of course, no one has ever dealt with time correctly in an AI program, and there is reason to believe that doing it would change everything." [McDermott82]

Since then the relevance of temporal information for diagnosis and other tasks has been widely accepted; yet there are comparatively few existing systems which have made progress towards a representation of time as a dimension in its own right. In this chapter we shall describe several of these approaches; in particular we shall take a closer look at the different techniques they use for the representation of temporal data. The presentation will be divided into two parts: in the first section we are concerned exclusively with diagnosis and describe individual systems whereas in the second section we shall discuss other domains in which AI systems have contributed to a better understanding of the representation of time. We will conclude the chapter by stating how our own approach relates to the work reported.

2.1 The Treatment of Time in Existing Diagnostic Systems

Our exposition of diagnostic systems will be tightly focused on contrasting the various approaches and comparing them to the position taken in this thesis. Therefore, instead of dwelling on idiosyncrasies we will try to characterize each system's approach to temporal representation by answering the following key questions:

1) Is the representation declarative or procedural?

2) If it is declarative, which kind of formalism is used (first-order predicate logic, reified temporal logic [Shoham88], modal logic, ...)?

3) Does the representation provide for partial temporal information? If so, how?

4) Is the underlying time line quantitative (e.g. the reals) or qualitative (sequence of states, symbolic intervals, ...)?

5) Is the application of temporal knowledge observation-driven or hypothesis-driven?

2.1.1 VM and MED2

In his introductory survey [Puppe88] Frank Puppe describes VM [Fagan84] as one of the earliest expert systems operating in an inherently temporal domain. VM's task is to monitor iron lung patients; the system's input consists of time-stamped sensor readings at regular intervals of 2-10 minutes.

Time	11.30	...	12.19	12.20	12.30
Breathing frequency (minute^{-1})	9	...	10	9	9
Blood pressure (mm Hg)	150	...	153	154	141
...	...	...	...	...	...

Fig. 1 – Sample input for VM

The resulting value histories for the individual parameters are then evaluated using special temporal predicates. These predicates allow certain trends or envelopes of the measured values to be expressed declaratively. As an example, a rule can refer to the condition that the fluctuation of blood pressure has not exceeded 15 mm Hg during the last 20 minutes. The introduction of special temporal predicates represent a step in the direction of reified temporal logic [Shoham88], although the temporal and atemporal arguments to these predicates are not separated clearly. There are also no explicit axioms for manipulating formulae containing temporal predicates; the matching algorithm is quite primitive: it is completely observation-driven (all sensor input arrives automatically) and is rerun on the whole set of observations each time a new measurement occurs. For representing information the language provides only the temporal predicates; there is no way of defining new predicates for situations not anticipated in the original design of the language. All temporal relations in VM are exact in the sense that temporal formulae refer directly to the discrete time stamps of the measurements. By contrast it is not possible to refer to intervals of indeterminate length.

MED2 [Puppe87] is based on a similar time representation, although here it is not used to express the short-range evolution of value histories, but instead to provide continuity of the diagnostic process across more than one session. Many existing diagnostic expert systems are not able to take up an earlier case at a later point in time: they either cannot distinguish between the times of the different sessions and run into conflicts wherever a parameter is observed to have different values on two occasions, or they sidestep this problem by treating the second session as a separate new case thereby losing the opportunity to reason about and correlate data from both sessions at the same time. Because of their similarity MED2 inherits the limited expressiveness of its representation language from VM. The matching algorithm, however, has been replaced by an improved incremental algorithm which updates the dynamic knowledge base rather than reconstructing it from scratch for each session. Furthermore, since the intervals between updates are much longer in MED2 than in VM, techniques have been included to evaluate changes from one session to the next in the light of therapeutic effects.

2.1.2 Wetter's System for Ear Disorders

In his PhD thesis [Wetter84] Thomas Wetter describes the architecture of an expert system diagnosing ear disorders. The basic terminology of his system consists of symptoms, information vectors, tests and disorders (fig. 2).

Wetter distinguishes two temporal aspects:

1) The diagnostic process is in itself sequential, i.e. starting with an empty information vector a sequence of tests is performed each of which adds its result monotonically to the information vector. Each possible sequence of tests thus corresponds to a sequence of vectors until either a diagnosis is found or the tests are exhausted. This aspect is present in Wetter's system regardless of whether or not the formulae defining the disorders are time-dependent.

2.) Additionally, $\varphi^+(H)$ and $\varphi^-(H)$ may contain temporal references which make the definition of a disorder dependent on symptom values at different points in time.

Concept	Example
Each *symptom* s takes on values from a domain r_s.	Body temperature has domain [35.0;45.0].
Information vectors v are sets of all symptom values known at a given stage of the diagnostic process.	v = {body temperature = 38.3, inflammation = 1}
Tests add symptom values to information vectors for previously undetermined symptoms.	By asking the appropriate question we could determine whether the patient feels pain and add "pain = 0" or "pain = 1" to v.
Disorders H are defined by a pair of formulae $\varphi^+(H)$ and $\varphi^-(H)$ over symptom values. $\varphi^+(H)$ specifies when a disorder is considered proved, $\varphi^-(H)$ specifies when it is considered disproved. The two formulae are not necessarily exclusive nor do they cover the cross product of the symptom domains.	$\varphi^+(H)$ = body temperature > 38.0 $\vee$ inflammation = 1 $\varphi^-(H)$ = body temperature < 37.0 $\wedge$ pain = 0

Fig. 2 – Basic terminology of Wetter's system

Both temporal aspects are treated in the same way[6]. The diagnostic process is modelled in the form of an acyclic directed graph in which the nodes are labelled with information vectors and the edges are labelled with tests transforming one information vector into the other. This graph can be given a modal logic interpretation: a hypothesis is confirmed in a given world w (= information vector), if every path from w leads to a world in which $\varphi^+(H)$ holds, i.e. if $\Box_w\ \varphi^+(H)$. H is possible in w, if $\Diamond_w\ \varphi^+(H)$, and so on.

---

[6] This is interesting, because it makes Wetter's system an early example for the philosophy proposed in recent literature [Struss89], that the diagnostic *process* should be accessible to the diagnostic reasons on the same level as the domain knowledge.

Temporal information can be handled by superimposing a second family of modal logics on this basic model (fig. 3). Let τ be an ordered set of relevant temporal entities (points and/or intervals) and assume that for each $t \in \tau$ there is a world w_t. A world w_t is accessible from world w_s, if s directly precedes t in τ. Each world w_t is itself structured and contains an instance of the basic model discussed above.

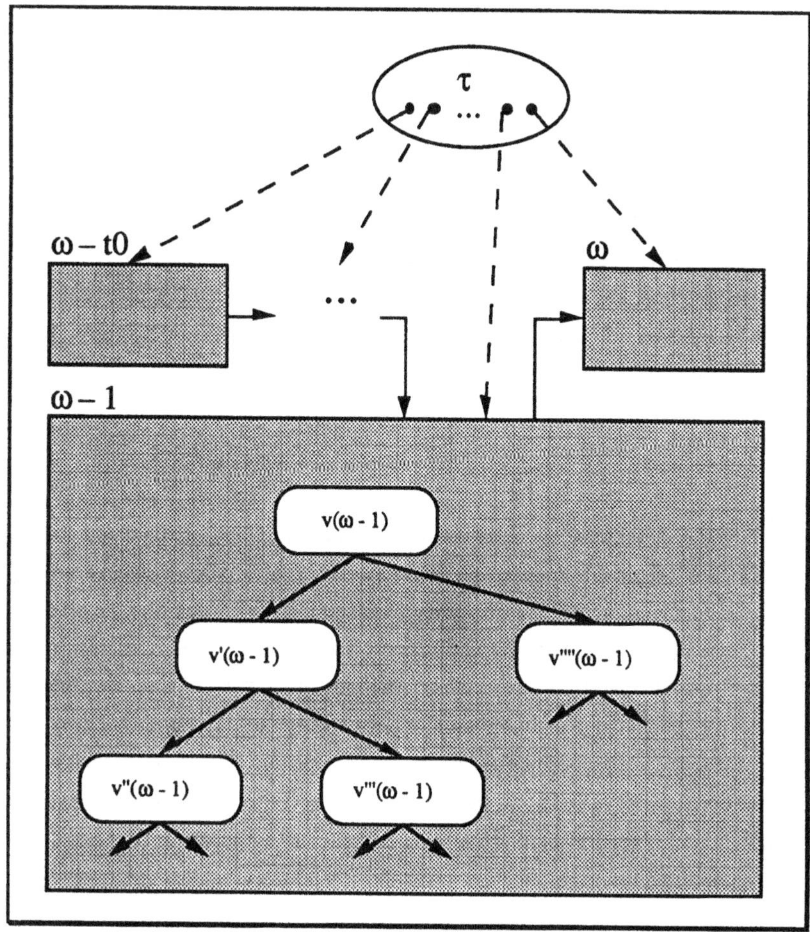

Fig. 3 – The two layers of modal logics in the temporal version of Wetter's system (ω denotes "now")

The basic terminology can be carried over to the temporal case without much added complexity. Disorders become modal formulae (with indices from τ), e.g.

$$\varphi^{+}(H) = \Box_{\{t \mid w\text{-}t \leq t_0\}} (s_1 - s_2) > a_{12} \;\wedge\; \Diamond_{\{t_1 \leq t_2\}} s_5 = 1$$

means that H is confirmed, if $(s_1 - s_2) > a_{12}$ throughout $[w\text{-}t_0; w]$ and $s_5 = 1$ at least once in $[t_1;t_2]$. Similarly the current status of a diagnosis is now a doubly modal formula. Wetter has worked out a set of theorems which can be used to manipulate doubly modal formulae and which permit the planning of further tests.

Wetter's system uses modal logic to explicitly represent temporal data in a declarative manner. The modal formulae are directly accessible by the reasoner and form the basis for hypothesis-driven[7] planning of observations. The granularity of the underlying time line is qualitative and can be varied according to the nature of the observed symptoms (by choosing an appropriate τ). Since the indices of the modal operators in the hypothesis formulae may be arbitrary complex, in principle every form of partial temporal information can be expressed. However, the resulting indices may become very complex. There is no indication of a reasoning system that is specially designed for the type of inferences likely to occur in such a system (beyond stating the axioms) and there is as yet no implementation of the temporally augmented architecture.

2.1.3 Alven

The main objective in John K. Tsotsos' work on the Alven system [Tsotsos85] was to demonstrate that diagnostic subtasks such as spatial, temporal and causal reasoning require a methodological framework (Tsotsos called it "knowledge organization") that goes beyond conventional rule-based systems and more closely resembles the elaborate organizational structures used by human experts. Tsotsos had chosen a very ambitious domain, left ventricular (LV) performance assessment from x-ray image sequences, which involves primarily reasoning about values and spatial relations changing over time. He was faced with the severe problem that LV dynamics was both critical for the interpretation of observations and not well understood theoretically. Mathematical modelling did not seem adequate and alternatives such as e.g. qualitative models were only just emerging at the time.

As a consequence, Alven could not be model-based and was therefore designed to integrate the spatio-temporal aspects into the associative paradigm. It is in this sense an intellectual predecessor of the MOLTKE system described in section 4.1 which also uses advanced knowledge organization principles to represent different kinds of associative diagnostic knowledge.

---

[7] except for an observation-driven initialization phase.

Of the various ramifications of Alven's knowledge base and control structure we are interested in the treatment of time-varying data only. Whenever a hypothesis is about to be refined, Alven must discriminate between all possible refinements of the current hypothesis. Discrimination involves the search for significant symptoms which can in turn proceed in one of several different modes (failure-directed, data-directed, model-

```
volume: VOLUME_V with
volume ← (vol of VOLUME_V with
  vol ← (minaxis.length @ now) ** 3
  default (117 @ m.systole.time_int.st,
          22 @ m.systole.time_int.et,
          83 @ m.diastole.rapid_fill.time_int.et,
          100 @ m.diastole.diastasis.time_int.et,
          117 @ m.diastole.atrial_fill.time_int.et)
  such that [
  volume @ m.diastole.time_int.et >= 97
          exception [TOO_LOW_EDV with volume ← volume],
  volume @ m.diastole.time_int.et <= 140
          exception [TOO_HIGH_EDV with volume ← volume],
  volume @ m.systole.time_int.et >= 20
          exception [TOO_LOW_ESV with volume ← volume],
  volume @ m.systole.time_int.et <= 27
          exception [TOO_HIGH_ESV with volume ← volume]],
  time_inst of VOLUME_V with time_inst ← now
```

Fig. 4 – An example for the definition of a time-varying quantity in Alven. Both default values and range conditions for exceptional situations are shown for different time points.

directed[8], and temporally directed). In the temporally directed search time-stamped observations are matched against descriptions of events and for each hypothesis events are predicted which are likely to occur next. The main difference from the approach

---

[8] "Model" refers here to an IS-A hierarchy of hypotheses rather than to a behavioral model.

taken in this thesis is the form in which temporal symptoms are specified. Since Alven cannot base observation planning on a situation description obtained from simulation, observation plans are hard-wired into procedures written in a kind of programming language. The language provides primitives for making observations at specific time points and constructs for the aggregation of subevents into events and expressing temporal precedence between events. There is a strong structural resemblance between Alven's language and the language used to write testing programs for hardware testing.

Basically, matching works as follows:

1) The testing program for the symptom to be verified is run on the set of existing observations to establish the current context.

2) If the program cannot proceed beyond a certain point because of missing data, the observations following that point in the program are predicted as the next events. If a measured value falls outside the predicted range, an exception occurs and the match fails with an annotation of what happened. Otherwise, the match succeeds.

In summary, Alven uses a procedural representation for hypothesis-driven matching. The elimination of a separate observation planning step based on a declarative situation description is particularly understandable, since Alven's domain contains complex interactions of spatial and temporal relations for which still no straightforward and computationally effective extensions of temporal calculi are known. The examples given in [Tsotsos85] do not allow a judgement about how partial information might be programmed into a testing procedure. All tests throughout the examples refer to fixed time points on a real time line.

2.1.4 SIDIA

The SIDIA project [Guckenbiehl89] is one of several research projects which aim to augment the popular GDE[9] paradigm for model-based diagnosis in various ways. SIDIA's application domain is a simple computer similar to a PDP-8 which is modelled on different levels ob abstraction. While on the higher levels the behavior is assumed to be static (or rather instantaneous), the temporal resolution of the lower levels is fine enough so that the efforts of set-up and hold times as well as gate delays become noticeable.

---

[9] For an overview of GDE see section 4.2.1.

Previous systems have dealt with these effects in a relatively coarse way; Guckenbiehl gives the example of a specification of the behavior of a JK-flipflop (fig. 5) that is typical of diagnostic systems whose time scale is on the order of the clock pulses:

J(t)	K(t)	Q(t+1)	~Q(t+1)
0	0	Q(t)	~Q(t)
0	1	0	1
1	0	1	0
1	1	~Q(t)	Q(t)

Fig. 5 – Discrete specification of a JK-flipflop

This specification is, however, inadequate, if the system has to reason about phenomena between clock pushes. In order to identify hazards or similar timing problems it has proved necessary in SIDIA to replace the discrete time line by a time line that is isomorphic to $\mathbb{R}$. In this model the set operation of the JK-flipflop which corresponds to the third row in fig. 5 can be described in greater detail taking into account both set-up and hold times:

```
∀t: if          T(CP = 0, before (t))
          and   T(CP =1, after (t))
          and   T(J = 1,  (t - 30, t + 5))
          and   T(K = 0,  (t - 30, t + 5))
    then        T(Q = 1,  after (t + 30))
          and   T(~Q = 0, after (t + 20))
```

Fig. 6 – More detailed specification of a JK-flipflop (CP is the clock pulse)

Most existing episode constraint propagation systems (e.g. TCP [Williams86], EP [Decker87], CONSAT [Güsgen/Fidelak88]) can only handle propagation rules that refer to *single* time points and apply these rules to all points in an interval in parallel.

Allowing the propagation rules themselves to refer to time *intervals* in both their antecedent and action part requires an extension of episode propagation that is described in [Guckenbiehl89]. The central idea is to treat the matching of the temporal terms exactly like the unification of the atemporal term[10]. Given two formulae

$$\Phi_1 : \forall t \in \text{tm: } T\ (\xi, \text{tsg}(t)) \qquad \text{(the data)}$$

and $$\Phi_2 : \forall t' \in \text{tm: } T\ (\xi', \text{tsg}'(t')) \qquad \text{(the pattern)}$$

where tm is a set of time points and tsg(t) and tsg'(t') are terms denoting time intervals relative to t, not only ξ and ξ' have to be unified; one also has to compute the "intersection" of the intervals denoted by tsg(t) and tsg'(t'), i.e. the interval during which the resolvent (the inferred episode) holds. As Guckenbiehl shows, this unifier can be computed in a straightforward manner for a wide variety of temporal terms. Most of these terms denote intervals of fixed length; notable exceptions are before(t) and after(t) which correspond to Allen interval relations. Their treatment is problematic because propagation may yield other terms denoting indefinite length intervals all of which have to be treated individually during unification. In summary, Guckenbiehl uses a reified temporal logic in the sense of [Shoham88] to declaratively specify temporally distributed events. Event patterns may refer to intervals on a continuous time line although in the present form there is only limited support for the expression of truly qualitative intervals. In SIDIA's domain this is justified because most effects have a precisely defined duration. [Guckenbiehl89] does not state clearly whether and how episode resolutions can be used to plan observations although there is an indication that matching takes places purely numerically with special provision only for "before(t)" and "after(t)".

---

10 Therefore the method is called *episode resolution*.

2.2 Time in Non-Diagnostic Systems

Up to now we have discussed the representation of temporal data only in a diagnostic setting. In fact, the task of relating temporally indexed observations to a hypothesis of what is happening is a generic task that occurs not only in diagnosis but also in a variety of other contexts, both in and outside AI.

In this section we will discuss some of these areas and show which aspects they have in common with the problem of dynamic symptoms in diagnosis and in which they differ.

2.2.1 Measurement Interpretation / Model-Based Monitoring

One of the basic tasks that a model-based reasoning system can solve is to explain measurements taken in a system in terms of a state sequence that both occurs in the system's envisionment and accounts for the observations. In terms of input and output measurement interpretation is the dual of temporal matching, the task studied in this thesis. The former works from a set of measurement sequences which have already been observed[11] towards a segment of behavior which gives rise to the observations, while the latter is given a hypothesis about the behavior and tries to establish it by proposing to observe a suitable measurement sequence.

Forbus contends ([Forbus86], [Forbus83]) that measurement interpretation is crucial for many applications of model-based reasoning to engineering tasks and indeed it is usually viewed as the core of e.g. model-based monitoring. Actually, however, this is true in the strict sense only as long as no a priori hypothesis has been formed about the system's behavior and there is no hypothesis-dependent choice of measurements. More recent papers (cf. [Doyle et al. 89], [Dvorak/Kuipers89]) show that this quickly becomes infeasible if the space of potential measurements (as e.g. the number of sensors in a complex technical system like a space station) and/or the number of alternate behavioral modes (normal operation and several fault modes) becomes very large. A logical strategy to tame this complexity problem is to make the monitoring process more informed: as soon as a hypothesis about the cause of a deviation is

---

[11] or which are being observed without guidance by the system.

formed, it can be used to plan further measurements according to the methods described in parts 3 and 4 of this thesis.

In fact, the similarity between measurement interpretation and temporal matching goes deeper than the input-output level. Many problems encountered in the course of solving one problem reappear in the order in a different guise. An interesting example is discussed in [Forbus86]: an easy special case of measurement interpretation occurs when

1) there is a non-zero duration *st* such that all state changes in the system's envisionment are at least *st* apart and

2) the largest gap between any two measurements for the same quantity is less than *st* (such a measurement sequence is called *close*).

If the input sequence is close, we are guaranteed to have directly observed every state that will be included in the explanation. In contrast to this, explaining a sparse measurement sequence involves guessing "hidden" intermediate states; it can be easily seen that the guessing process bears the chance of a combinatorial explosion, hence it is advantageous to start with a close measurement sequence whenever possible.

In section 3.5 we will discuss the limitations of verifying a segment of continuous behavior using only discrete measurements. The most striking limitation, of course, is the inability to cover a continuous interval with measurements and therefore the impossibility to infer the behavior between individual measurements. Not surprisingly, the compromise we will adopt is to postulate the equivalent of a "minimal persistence duration" and to restrict ourselves to measurement sequences which are close.

2.2.2 Hardware Testing

Although hardware testing and the fault diagnosis of electronic circuits (as exemplified in SIDIA) have much in common, they serve different purposes which to a certain degree influence the techniques used. Diagnosis *presupposes* that the device considered displays an "abnormal" behavior and constructs a causal explanation for the discrepancy which may be a known or even an unknown fault. In contrast, hardware testing verifies that *none* of a set of known faults is present in a given artifact; the primary distinction is "normal/abnormal" whereas the identification of the particular cause in the abnormal case is of secondary importance. "Abnormal" and "normal" are, of course, relative terms and have to be interpreted teleologically, i.e. with respect to the intended function of the artifact. The distinction between normality and abnormality is even less significant on the methodological level: both types of behavior can be modelled in exactly the same way; any technique that has been developed in order to detect one type can also be employed to detect the other. In this sense much of the work in hardware testing since the time of the venerable D-algorithm [Roth et al. 67] has a direct impact on diagnosis although this has been somewhat overlooked in the past. Naturally, many algorithms rely heavily on simplifying assumptions (e.g. digital systems, only static behavior, no feedback) which are justified for a wide class of electronic equipment but make it hard to extend the algorithms to a broader range of application domains. Nevertheless the basic steps in test generation are universal and efforts have been made recently to overcome the limitations (examples for this tendency can be found in [Shirley88] and [BeckerU89]).

2.2.3 Event Recognition in the High-Level Analysis of Image Sequences

One of the last stages in the analysis of image sequences is finding a compact description of what happens in the sequence in terms of conceptional units called events[12]. While the previous phases have identified objects and their trajectories and thus have abstracted from the pixel level, high-level analysis searches for patterns in the scene description which correspond to certain prespecified events and generates concise natural language descriptions for them.

High-level sequence analysis has been one of the objectives in both the NAOS [Neumann84] and VITRA [Rist et al. 87] projects. The approach taken in NAOS requires that a description of the *complete* image sequence be given as input to the high-level analysis module; VITRA's analysis is closer in spirit to the type of event recognition studied in part two of this thesis in that image sequences are processed *incrementally*.

Events are specified as pairs of a set of agent variables and an event schema[13]. Basically, event schemas are finite automata whose edges are labeled with predications over the agent variables. The schemas can be thought of as translations of concept definitions in Allen's temporal logic. [Rist et al. 87] gives the example of the concept "Pass_to_running_player" from soccer which could be defined as follows:

---

[12] Events can be interpreted in the same sense as in [McDermott82].

[13] I have translated the terminology from the German paper [Rist et al. 87].

occur ($tint_1$ (Pass_to_running_player))

⇔

∃ $tint_2$, Player1, Player2:

DURING ($tint_1$ $tint_2$)

∧ OCCUR ($tint_2$ (Run Player2))

∧ OCCUR ($tint_1$ (Pass Player1 Ball Player2)).

The end points of all intervals in the definition are then projected onto a time line and each segment between two consecutive end points is translated into an arc of the automaton which is labelled with the set of predications true in the segment. Consider e.g. the situation of two overlapping intervals in fig. 7 (a) which is translated into the schema shown in fig. 7 (b)

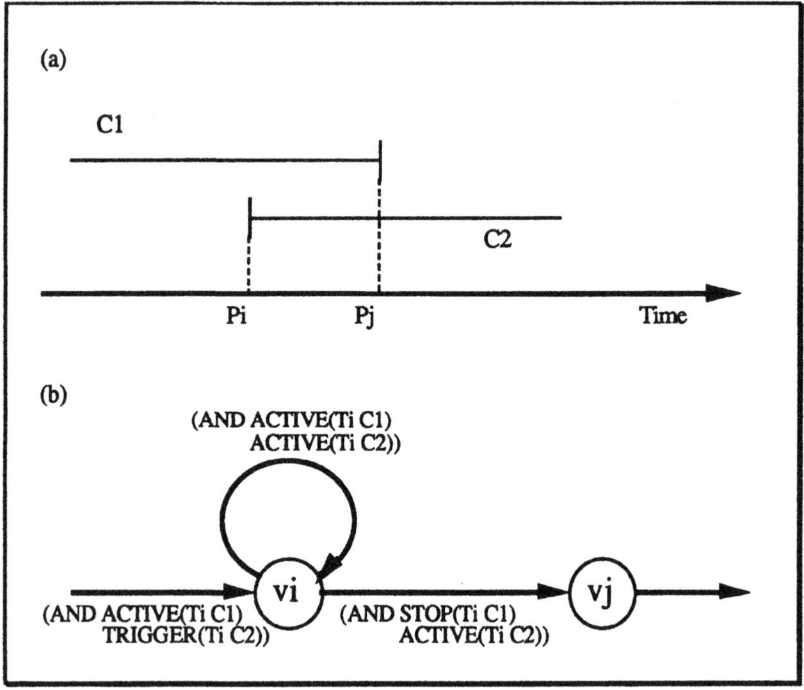

Fig. 7 – (a) Overlap between concepts C1 and C2 and (b) corresponding partial event schema (from [Rist et al. 87])

In the same manner the original example of "Pass_to_running_player" yields the event schema in fig. 8.

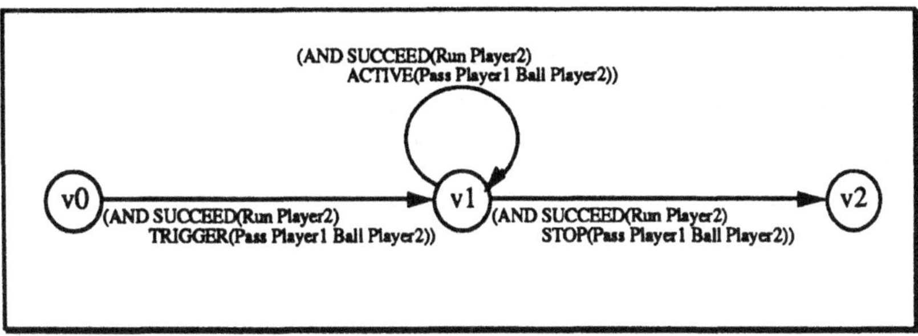

Fig. 8 – Complete event schema for the event "Pass_to_running_player" (from [Rist et al. 87])

Recognizing an event in an image sequence then means traversing a path from the starting node of the event schema to an end node in a 1–1 correspondence between images and areas where the arc labels have to be consistent with the descriptions of their individual images.

The representation of events in VITRA differs from the definition of situations in this thesis[14] in two points. Firstly, VITRA admits arbitrary formulae as arc labels making the language very flexible at the cost of complicating the matching step, i.e. determining expected events in a given situation and deciding whether the observed data (the images) are compatible with them. In technical diagnosis situations are more uniform than in VITRA's or NAOS' domain so that we can restrict the language without great loss in expressiveness. Secondly, the translation of concept definitions in Allen's logic into event schemata is problematic: in principle the projection of interval end points onto a time line is possible only if the interval relations are singletons. Disjunctive interval relations, however, lead to more than one projection and hence to more than one event schema all of which would have to be processed in parallel (or, sequentially with backtracking). Since the translation process has not been automated in VITRA, this multiplication of schemas would be both tedious and a source of errors. The matching algorithm in part 3 is an alternative that avoids duplicate schemas because it operates directly on the Allen-style representation and efficiently computes expected observations etc. *only along* the path actually taken during recognition.

---

[14] see section 3.2.

2.2.4 Time Series Analysis

Another form of relating observations to hypotheses about what is going on is the subject matter of statistical time series analysis. It is concerned with the analysis of data which are not independent but serially correlated and where the relations between consecutive observations are of interest. Box and Jenkins [Box/Jenkins70] developed a very popular method of time series analysis which proceeds in an iterative cycle of identification, estimation and testing that seemingly resembles the cycle of hypothesis formation, discrimination and verification in diagnosis. As mentioned in the introduction, the similarity breaks down when the respective time scales are considered. Of course, there are phenomena in technical diagnosis which can be modelled as parameters undergoing long-term changes. A good example is the effect of tool wear on product quality where statistical methods can help prevent unacceptable deterioration. Another example is the analysis of high-frequency phenomena (such as vibration) where individual observations of the signal are not practical and statistical aggregations are used as substitutes. In contrast, the temporal aspects that we are interested in refer to phenomena on a time scale where statistical abstractions are not meaningfully applicable. In these situations signals can be observed in each qualitative state and the number of observations needed for discrimination is orders of magnitude smaller than is necessary for significant statistical results.

2.3 What is the Relation to our Work?

The significance of the work reported in this thesis in comparison to the diagnostic systems discussed before can be understood best by going through the same set of questions. Fig. 9 summarizes the answers for VM / MED2, Wetter's system, Alven, SIDIA and also for our own approach.

As one can see, our system differs markedly from VM / MED2 in that it is primarily hypothesis-driven and therefore more efficient in terms of effort spent on observations. It also differs from Alven in the use of a declarative representation of TDSs which is better suited as a platform for knowledge acquisition from both qualitative simulation results and heuristic knowledge of human experts. There are several resemblances between our system and the spirit of Wetter's temporal extension; the latter has never been implemented, however, and we suspect that the attempt would turn out to be difficult because the representation language is very strong and no efficient algorithms for strategic planning of observations are to be expected. The alternative we propose is

	declarative vs. procedural representation	formalism	partial information expressible?	qualitative or discrete time line?	matching observation-driven or hypothesis-driven?
VM / MED2	declarative	fixed set of temporal predicates	–	discrete	O
Wetter	declarative	modal logic	+	discrete	H
Alven	procedural	"test programs"	?	discrete	H
SIDIA	declarative	temporal constraints	(+)	discrete	O / H
our work	declarative	TDS repr. based on Allen's temp. logic	+	qualitative *and* discrete	H

Fig. 9 – Summary of system characteristics

to use a comparatively modest representation formalism which nevertheless covers the range of phenomena encountered in technical domains fairly well and supports efficient algorithms. SIDIA is perhaps the closest relative of our system – at least if we compare it to the use of TDSs in a model-based diagnostic setting proposed in section 4.2. Our representation language provides more flexibility in expressing partial information whereas SIDIA's is superior at capturing quantitative information (e.g. durations). This is primarily a consequence of our choosing a qualitative time interval based representation instead of a quantitative, discrete time line. Furthermore we view it as an advantage of our approach that it is equally suited for an associative/heuristic or a model-based (or rather simulation-based) diagnostic system. In this sense we claim that it is a useful paradigm-independent building block in the construction of expert systems for technical diagnosis.

As one can see, our system differs markedly from VM / MED2 in that it is primarily hypothesis-driven and therefore more efficient in terms of effort spent on observations. It also differs from Alven in the use of a declarative representation of TDSs which is better suited as a platform for knowledge acquisition from both qualitative simulation results and heuristic knowledge of human experts. There are several resemblances between our system and the spirit of Wetter's temporal extension; the latter has never been implemented, however, and we suspect that the attempt would turn out to be difficult because the representation language is very strong and no efficient algorithms for strategic planning of observations are to be expected. The alternative we propose is to use a comparatively modest representation formalism which nevertheless covers the range of phenomena encountered in technical domains fairly well and supports efficient algorithms. SIDIA is perhaps the closest relative of our system – at least if we compare it to the use of TDSs in a model-based diagnostic setting proposed in section 4.2. Our representation language provides more flexibility in expressing partial information whereas SIDIA's is superior at capturing quantitative information (e.g. durations). This is primarily a consequence of our choosing a qualitative time interval based representation instead of a quantitative, discrete time line. Furthermore we view it as an advantage of our approach that it is equally suited for an associative/heuristic or a model-based (or rather simulation-based) diagnostic system. In this sense we claim that it is a useful paradigm-independent building block in the construction of expert systems for technical diagnosis.

3 Temporal Matching

According to the distinction made in the introduction we will assume throughout this chapter that the relevant TDSs in the domain under consideration are already known. This is not a bad magician's trick, but a common situation in the construction of an associative expert system in general: except for relatively superficial plausibility checks the correctness of the expert knowledge is taken for granted. Instead of trying to find a causal justification for a rule or a fact the work focuses on representing the available expertise and making it operational, i.e. designing inference mechanisms for the chosen representation format. Separating the usage of expertise from the various ways of acquiring it is not just any way of decomposing the problem; finding a solution to the representation and reasoning subproblem is useful, even if there were no way of (partially) automating the knowledge acquisition process. In fact, this is how we approached the problem of TDSs in the MOLTKE project (see section 4.1). It seems reasonable, therefore, to describe this solution first; we will return to the subproblem of deriving descriptions TDSs from predicted behaviors in the context of model-based diagnosis (section 4.2).

This chapter is organized in the following manner: following an intuitive statement of the task we will discuss how the inputs - TDS descriptions and measurement sequences - can be represented. We will point out advantages and limitations of our approach in terms of expressiveness and computational complexity. The two dimensions are not independent of each other: in general, the more universal the representation language, the more expensive the operations on the language will be. We will strike a balance between the conflicting interests presenting a restricted language for temporal relations which has favorable complexity properties without sacrificing too much expressive power to be useful.

The next step is to define properties which a definition of temporal matching should have. Again it turns out that these properties in their ideal form conflict with some practical limitations forcing us to find a weaker version. We will prove that there is a definition of matching that corresponds to this set of weaker properties. Finally, we operationalize the definitions by describing an algorithm that incrementally plans and matches measurements in order to verify the occurrence of a TDS.

Throughout this discussion we will restrict ourselves to TDSs that can be described in a purely qualitative fashion. Our experiences in the MOLKTE project indicate, however, that at least a limited form of quantitative information is needed to specify realistic TDSs. Since the presentation of the representation formalism, the definition of temporal matching and the algorithm would unnecessarily be cluttered by this extension we postpone its treatment until the last section of the chapter where we will characterize the type of additional quantitative information and show how our approach to temporal matching can be augmented in an adequate way.

3.1 Informal Statement of Problem

Let us begin with an informal explanation of what we want our system to be able to do. We will do this by walking through a simple example (fig. 10) from MOLTKE's domain identifying subtasks and knowledge sources as we go. We will return to the example later in this chapter where it will serve to illustrate the formal definitions.

> One possible cause for an undefined position of the tool magazine is a faulty limit switch. This cause can be ruled out if the status registers IN29 and IN30 of the CNC control system show the following behavior: at the beginning both registers contain the value 1. Then IN29 drops to 0, followed by IN30. Finally, both return to their original values in the reverse order.

Fig. 10 – An example of a TDS

If we separate the information defining the TDS from the information about the diagnosis associated with it (which does not concern us here), we obtain the graphic representation in fig. 11.

We can make a number of observations in this example:

- The specification of the sequence of events in the TDS does not contain any quantitative references (such as e.g. durations). Only the relative temporal order of value changes is relevant. The simultaneous start and end points of the two curves is insignificant, insofar as they are not mentioned in the textual version.

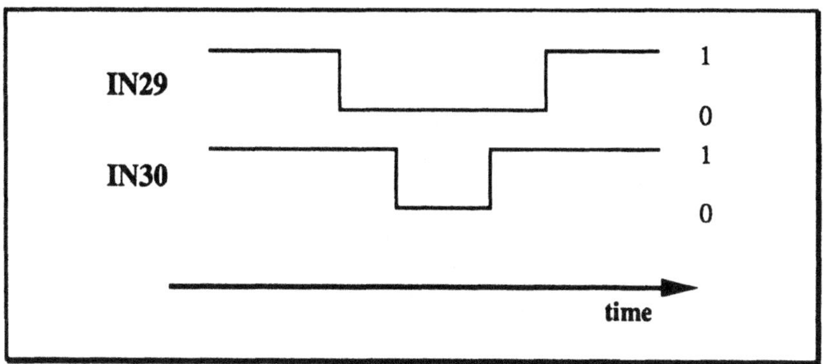

Fig. 11 – Graphic representation of the TDS

- Obviously, detecting an occurrence requires more than one measurement for each quantity and the temporal order of measurements is significant.

- The graphical representation is very intuitive and natural, but using it for detecting an occurrence of the TDS requires the cooperation of a human. Many diagnostic systems adopt this solution and shift the burden of choosing the "correct" measurements and processing the raw data to the user of the expert system. They treat TDSs as unstructured entities which are either confirmed or rejected as a whole and thus completely eliminate the time dependency.

In MOLTKE we take a different approach and represent time explicitly, because

- only a part of the measurements is carried out by humans; the rest of the data comes from sensors, and

- for bigger TDSs which contain many alternatives, planning observations economically is non-trivial and error-prone, if done by humans.

For these reasons we wish to let the diagnostic system plan measurements, predict observed values and match the actual observations against the predictions. A rough sketch of the black box specification of temporal matching is shown in fig. 12.

Trying to manage without human assistance (except for performing the measurement actions for some of the quantities) immediately poses a problem: the graphic representation of a TDS (or the verbal formulation, for that matter) is not nearly as amenable to algorithmic manipulation as it is to interpretation by a human. Our first step

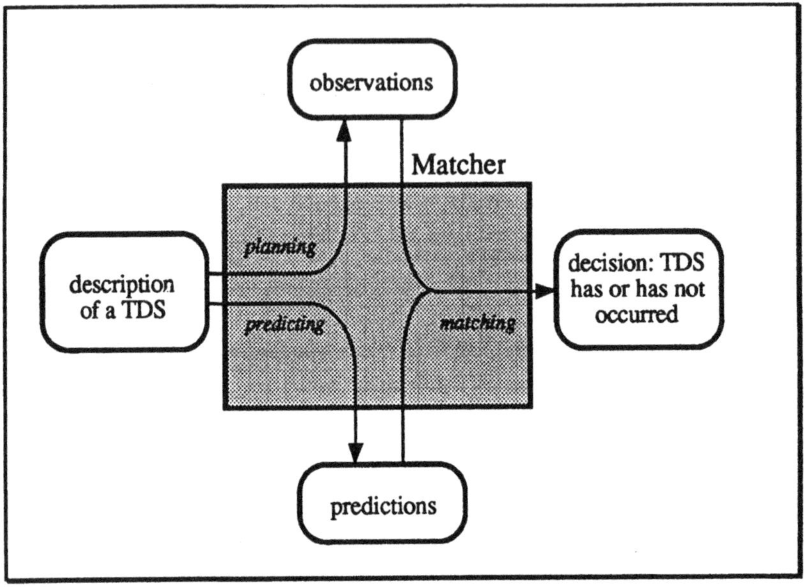

Fig. 12 – High-level specification of temporal matching (left to right)

therefore will be to look for an alternative representation format that better supports the operations we want to carry out on TDSs (see section 3.5).

Trying to manage without human assistance (except for performing the measurement actions for some of the quantities) immediately poses a problem: the graphic representation of a TDS (or the verbal formulation, for that matter) is not nearly as amenable to algorithmic manipulation as it is to interpretation by a human. Our first step therefore will be to look for an alternative representation format that better supports the operations we want to carry out on TDSs (see section 3.5).

The second important design parameter is the type of measurements available. Several possible scenarios are conceivable depending on equipment, the skill level of the person performing the individual measurement actions and other factors. It is easy to see that a situation where the number of significant parameters is small enough so that all of them can be monitored continuously imposes different demands on both representation and algorithms than a situation with sparse measurements would do. Intermediate cases can be imagined, too: in large-scale monitoring applications it may be possible to measure all quantities continuously, but due to the sheer volume of data there is not enough time to process all observations. Here the interpretation process must be focused using the current hypothesis about what is happening as a guidance. In this thesis we consider exclusively the scenario that we found in MOLTKE's application domain, which is fairly typical of technical diagnosis situations other than process control: although some machine parameters can be measured by sensors none of them is continuously monitored. Consequently, each individual measurement has to be explicitly planned and carried out. There are two axioms about measurements in MOLTKE which are central to our discussion:

M1) Measurements are instantaneous, i.e. quantities are measured at single points in time and not over intervals.

M2) No two measurements can be made exactly at the same time point. Due to the fact that in MOLTKE measurement suggestions always refer to only one quantity, two observations of different quantities are inevitably set apart in time by a minimal lag.

At the end of the thesis we will review which parts of our framework are sufficiently general to be used in other scenarios.

Suppose now that we want to detect an occurrence of the TDS using a sequence of instantaneous measurements. First of all, we have to define what we mean by "occurrence", since up to now we have talked about TDSs in the sense of abstract representatives for an infinite collection of potential events in the "real world". We must then decide under which circumstances we accept a measurement sequence as evidence for an occurrence of the TDS. Here a fundamental problem arises: since we cannot cover an interval with finitely many instantaneous measurements, we cannot possibly know what has happened between measurements. Nevertheless, measurement sequences do provide at least partial evidence: if e.g. IN29 is observed to be 1 even after IN30 has dropped to 0, what is going on cannot be an occurrence of our example TDS - unless, that is, our measurement sequence is very sparse indeed. In section 3.5 we will propose a solution to the dilemma which strikes a balance between additional assumptions about "real" occurrences of a TDS and the pragmatic goal to minimize the number of measurements.

But being able to decide a posteriori whether a given measurement sequence indicates an occurrence does not help much unless we have a way of effectively constructing such a sequence. What we need is a general matching algorithm which, given the TDS description and the measurements taken so far, proposes the next quantity to be measured and predicts its expected value. Suppose that we are in the process of observing an occurrence of our example TDS and we have already made the observations m_1, m_2, m_3 shown in fig. 13.

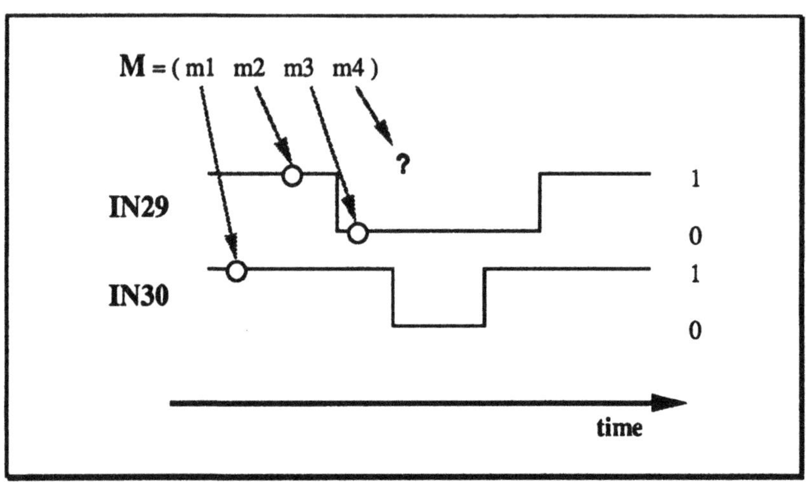

Fig. 13 – Planning measurements depending on previous observations

In this context the planning algorithm should come up with the suggestion "measure IN30 and expect 1" (shown as m_4), because this would be positive evidence for the 1-0-transitions being in the correct order.

Complementary to the planning step the temporal matching algorithm has to decide whether the actual observed value is compatible with the TDS or not. Not every deviation from the predicted value is automatically a contradiction. Consider again fig. 13 and assume that we have just suggested measurement m_3 with an expected value of IN29 = 0. If we proceeded to observe IN29 = 1 instead, nothing would be wrong: we might simply not have waited long enough for the transition to occur, so instead of reporting a failure we should try again. Deviations can also be the result of waiting too long: if we measured IN30 = 0 in response to the suggestion of m_4, we would have just missed the last chance of verifying the temporal sequence of the two 1-0-transitions. Again, the result of the matching process is not failure in the strict sense, but rather a state of "still-unknown-unless-we-start-all-over". The general idea for a temporal matching algorithm is to start with the TDS description and an empty measurement sequence and iterate through the planning and matching steps until either a contradiction is encountered or we arrive at a measurement sequence which satisfies the definition of matching derived in section 3.5. We will develop such an algorithm in section 3.6.

3.2 Specification of TDSs

3.2.1 A First Approach

The design of a representation formalism for TDSs depends on a number of factors:

- How are TDSs specified by experts? As long as TDS descriptions are compiled manually, we depend on experts' statements as our only knowledge source. The representation formalism should bear some ontological resemblance to the terminology of the experts; the closer the gap, the fewer the errors are likely to be made in the knowledge acquisition process.
- If on the other hand we want to use temporal matching in the context of model-based diagnosis, it must be possible to generate specifications of TDSs by simulating a model of the machine. In this case it is necessary to take the output format of qualitative simulation programs into consideration, because using a segment of the predicted machine behavior as the description of a TDS works best when there is a straightforward mapping between the two representation formats.
- Even more important, the representation formalism must support the operations that we will perform during the matching process itself. It is problematical to retrace the complex coevolution of representation and reasoning mechanism in a documentation with a linear organization. We will approximate this design process by anticipating some necessary algorithmic properties, although we have to postpone their justification until section 3.5.

All three requirements are independent of each other, so there is no reason to expect that a single representation formalism would be optimal for all of them. Instead of finding a common solution that is equally suboptimal in all three respects, we have developed three formats (fig. 14).

The internal representation is designed so that the matching process can operate on it. It is independent of the origin of the TDS and marks the boundary between knowledge acquisition/test generation and temporal matching. In this section we will be concerned with the internal representation only. The external syntax used by the developer of an associative diagnostic expert system to formulate TDS descriptions is covered in greater detail in section 4.1.2.1. The translation of simulation results into the internal representation for model-based diagnosis will be described in section 4.2.3.1.

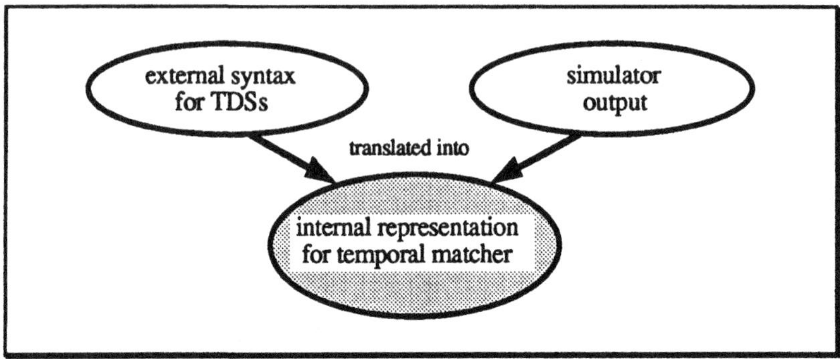

Fig. 14 - TDS description languages

Most of the basic ingredients of TDSs can already be found in the small introductory example. A TDS is characterized by one or more quantities taking on different values over a period of time. Not only the particular value sequence counts, but the temporal order of the two transitions is also significant. Since we can only measure a finite number of times, TDSs with continuously changing quantities are beyond our measuring techniques. We therefore require that all quantities are either discrete by nature or have been replaced by qualitative approximations by imposing an order-preserving equivalence relation on their values. Furthermore, quantities may change their values only finitely often during any finite interval. In summary, the "behavior" of a quantity over a period of time may be specified as a linear chain of intervals of maximal extent during each of which the quantity has a constant value. We will call each of these interval-value-pairs an *episode* a chain of episodes is called a *value history*. For TDSs the intervals are symbolic entities of non-zero extent in the sense of [Allen/Hayes85]; in section 3.3 we will use the same construction for the definition of "occurrences of TDSs" except that in this case the intervals are real intervals on the global time line.

A recurring question in temporal reasoning about episodes is whether the episode intervals should be open, closed or something in between, i.e. what happens at the end points? If all intervals are closed, we get a contradiction, because a quantity cannot take on two different values at the same time. Conversely, if all intervals are open, the value is undefined for the common end point of two consecutive intervals. For our purposes we assume all intervals to be closed. We do not, however, view two episodes overlapping in one time point as a contradiction but rather as a disjunction: the quantity may fake on either of the adjoining values and – for the sake of specifying the TDS – we do not care which (fig. 15).

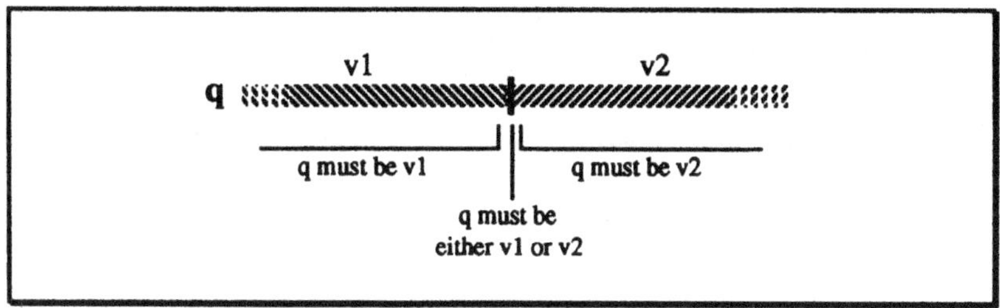

Fig. 15 - A solution to the end point paradox for histories in TDS descriptions

TDSs normally contain more than one value history. Depending on whether we view the temporal dimension or the distinction between the individual quantities as primary, we can organize a TDS description in either of two ways.

In the *state-oriented approach* we squash all value histories together to obtain a single tuple-valued history. This global history is constructed by conceptually projecting all histories - in particular all episode end points - onto a common time line. We thus obtain a chain of episodes during each of which *all* quantities remain constant. Each episode can therefore be interpreted as the projection of a global machine state onto the tuple of quantities involved in the TDS.

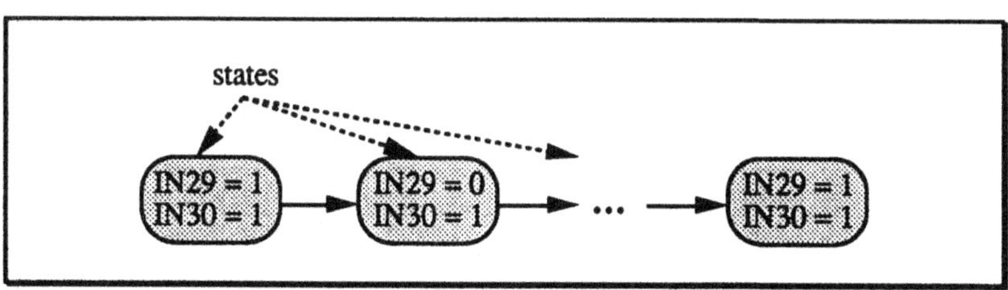

Fig. 16 - State-oriented representation of the example TDS

On the other hand, in the *quantity-oriented approach* we describe a TDS as a collection of individual histories - one for each quantity. Of course, a TDS is more than just that: in addition we need a way of expressing the constraints on the relative position of the individual histories w.r.t. each other; in fact, in our example TDS this is the salient feature.

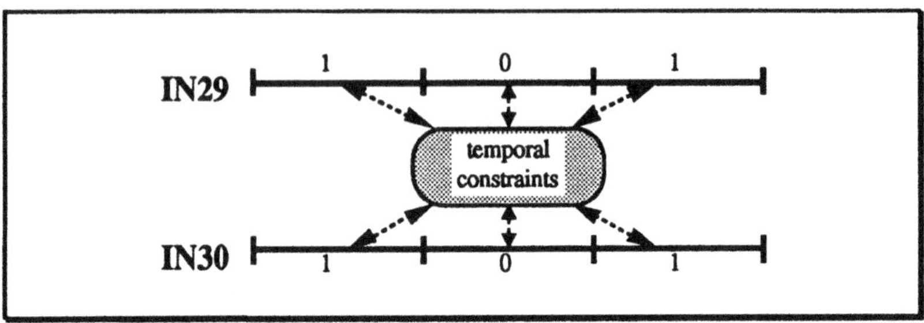

Fig. 17 - Quantity-oriented representation of the example TDS

Although, on the face of the matter, the state-oriented approach seems more natural, there is reason to study the quantity-oriented approach in greater detail. Contrary to our running example in some TDSs the temporal relations between the episodes are only partially specified. In addition to the imprecision caused by the qualitative nature of the intervals, it may be convenient to specify that e.g. two episode intervals have a non-

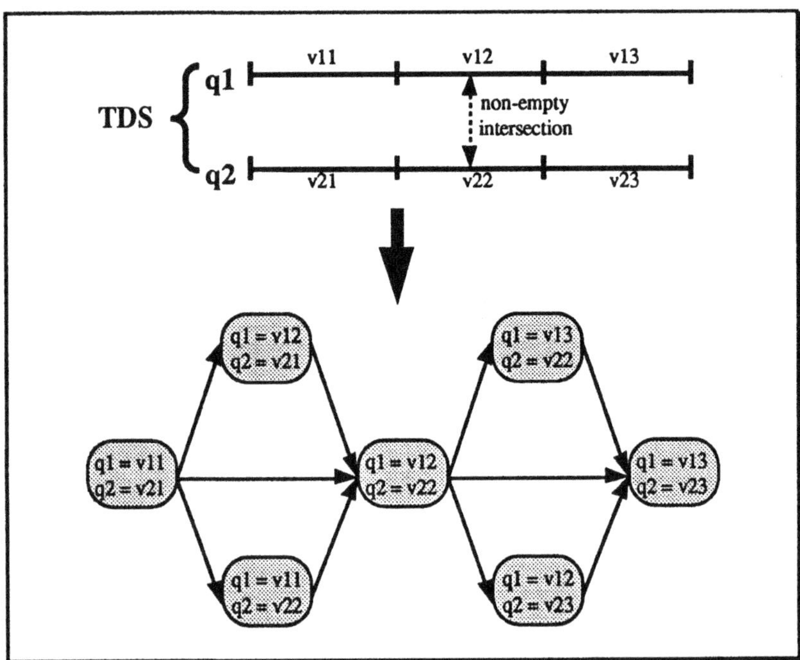

Fig. 18 - Incorporating partial temporal information into the state-oriented approach leads to a large graph of state sequences

empty intersection without fixing a particular ordering on the end points. If we try to incorporate this kind of partial temporal information into the state-oriented approach, we cannot find a unique state sequence any more (fig. 18).

The problem arises from the particular construction principle for state sequences so that we can hope to avoid it in the quantity-oriented approach.

In this section we have seen what TDS descriptions are made up of: A TDS is described by

- a set of quantities,
- a value history for each quantity,
- "some means" of specifying the temporal relations between the episodes of different histories.

We are not yet in the position to give a more formal definition; we still do not know which approach (state-oriented or quantity-oriented) to prefer and how temporal relations can be expressed at all in the quantity-oriented approach. This last question holds considerable interest in itself and we will devote the next three sections to it, taking up the matter of defining the representation formalism for TDSs again in section 3.2.6.

3.2.2 Interval Relations

If we are talking about specifying the relative positions of the temporal histories in a TDS description, we mean the temporal relations between their constituent episodes, or more precisely, between the intervals corresponding to the episodes. To this day there is a single formalization of interval relations which - together with several descendants - is by far the most popular in the AI community. Related to previous work outside AI (see e.g. [Bruce72]) it was proposed in [Allen83] as a means of representing and resolving temporal references in natural language utterances. Since then it has been applied to a multitude of other AI problems, such as planning, discourse representation and qualitative reasoning. It was further used as the underpinnings of a general-purpose common-sense temporal logic [Allen/Hayes85]. The popularity of Allen's interval calculus can be attributed to the rare coincidence that it is not only conceptually straightforward but it also allows computations to be performed on the interval relations (hence "calculus") which correspond to making simple inferences about interdependent intervals. For some problems there are even polynomial algorithms making Allen's calculus extremely attractive for AI systems that need to explicitly keep track of temporal relations without being swamped by complexity problems.

It therefore makes sense to take Allen's interval calculus as the starting point for finding an adequate representation of temporal relations in TDSs. On the flip side, though, Allen interval relations cannot express *all* interesting temporal relations that one can think of (see section 3.2.4), and one might be tempted to introduce further types of relations to cover these phenomena, too. However, small as Allen's relation algebra is, there are already many problems that can be posed in it which have been proven to be NP-complete.

In this section we will show how to strike a practical balance between expressive power and complexity. We start with a brief overview of Allen's interval calculus summarizing some important complexity results. We then show some examples of common situations that Allen relations fail to capture. On the other hand, among the theoretically possible relations there are many which are never used in TDS descriptions. This observation leads us to the characterization of convex relations which have favorable complexity properties, but are nevertheless sufficiently expressive for our purposes.

3.2.3 Allen's Interval Calculus

If the relation between two time intervals is characterized by the temporal order on their end points, we get 13 *primitive interval relations* which can be arranged in six pairs of mutually inverse relations plus identity. Fig. 19 gives a graphical definition of the 13 relations along with their common abbreviations.

Partial information about the relative position of two intervals can be expressed as a disjunction of primitive relations, e.g. we may express that A is a proper subinterval of B by stating

$$A\,s\,B \;\vee\; A\,d\,B \;\vee\; A\,f\,B$$

which is commonly abbreviated to A {s, d, f} B, or graphically

(A) —{s, d, f}→ (B)

Absence of any information about the temporal relation between two intervals can be expressed as the disjunction of all 13 primitive relations, A {<, >, m, mi, d, di, s, si, f, fi, o, oi, = } B, which we denote by A UNCONSTRAINED B. Information about

Relation	Symbol	Example
A before B	A < B	A
B after A	B > A	B
A meets B	A m B	A
B met by A	B mi A	B
A overlaps B	A o B	A
B overlapped by A	B oi A	B
A during B	A d B	A
B contains A	B di A	B
A starts B	A s B	A
B started by A	B si A	B
A finishes B	A f B	A
B finished by A	B fi A	B
A equals B	A = B	A
		B

Fig. 19 – The 13 primitive Allen relations

N>2 intervals is represented as the collection of all N^2 pairwise relations[15] which is called a *time net*; there is no way of directly expressing a relation over more than two intervals. Because of the central role time nets will play in the following discussion, we define them formally.

DEFINITION 1: A *time net* ⟨V,C⟩ is a complete labeled graph where

- V is a set of intervals (the vertices of the graph),
- C is the labeling function

$$C:\ V \times V \rightarrow \text{Power(UNCONSTRAINED)}$$

$$C:\ A, B \mapsto C_{A,B}.$$

DEFINITION 2: A time net ⟨V,C⟩ is called

(i) ***singleton***, iff $\forall$ A,B $\in$ V: $|\, C_{A,B} \,| = 1$;

(ii) ***unconstrained***, iff $\forall$ A,B $\in$ V: $C_{A,B}$ = UNCONSTRAINED.

---

[15] counting also the relation I = I, for each interval I.

To abbreviate notation we denote a time net of the form $\langle\{A,B\}, C\rangle$, $C_{A,B} = R$, as "A R B".

As the example in fig. 20 shows, the relations specified by C are not independent: of course, $C_{A,B}$ must be the inverse of $C_{B,A}$. More importantly, two relations may be composed to yield additional information about a third one.

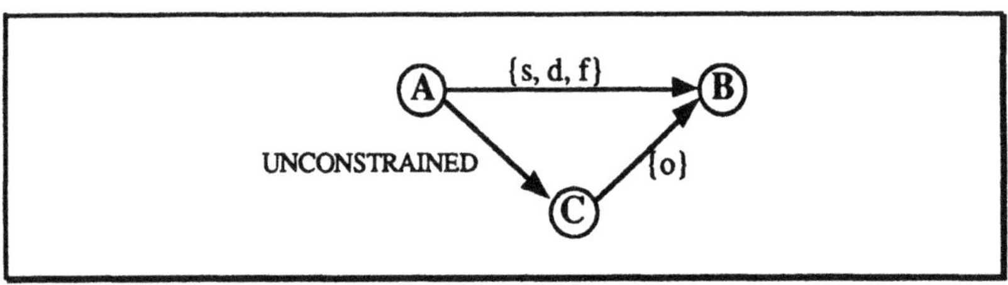

Fig. 20 – A time net for three intervals

Composition of primitive relations is defined by taking their transitive product: if we know that e.g. A {m} B and B {m} C, then necessarily A {<} C. The complete composition table is given in [Allen83]. Composition of disjunctive relations is defined as taking the union of the pairwise composition of their elements. In this way we can restrict the labels in a time net to more definite disjunctions; e.g. in fig. 20 by composing A {s, d, f} B and B {oi} C we can constrain A UNCONSTRAINED C to A {>, oi, mi, d, f} C. If later we get the additional information A {>} C, we can in turn infer A {d, f} B and so on. This style of reasoning is embodied in the general path-consistency algorithm which compares the direct relation between any two intervals with the result of composition along any path from one interval to the other. Allen's propagation algorithm [Allen83] is a special case where only paths of length 2 are considered; [van Beek89] describes an algorithm for length 3 paths.

In connection with time nets a number of satisfiability problems can be formulated some of which will be of importance later in this thesis. To be able to speak about satisfiability we first have to clarify the semantics of time nets $\langle V, C\rangle$. We will now inductively define the semantics in a model-theoretic way, starting with the special cases $|V| = 1$ and $|V| = 2$.

The models most commonly used for the intervals in Allen's calculus (A-intervals) are non-singleton intervals of $\mathbb{R}$ (R-intervals). We choose R-intervals to be closed. Generally models of time nets are functions of the form

$$D: V \to \text{Intervals}(\mathbb{R})$$
$$D: I \in V \mapsto [I^-; I^+],$$

i.e. mappings of the A-intervals to real intervals on the global time line.

An isolated A-interval has all R-intervals as models; formally:

DEFINITION 3:

$[\![\langle\{A\}, C\rangle]\!] := \{ D \mid D(A) = [A^- ; A^+], A^-, A^+ \in \mathbb{R}, A^- < A^+ \}$

(necessarily, C must specify $C_{A,A} = \{=\}$)

The next more difficult case is a time net for two A-intervals. If the relation between the two intervals is one of the primitive Allen relations, the models are defined according to fig. 19. For example the semantics of "A {m} B" can be stated as follows (the definitions for the other primitive relations are similar):

DEFINITION 3 (continued):

$[\![A \{m\} B]\!] := \{ D \mid D(A) = [A^- ; A^+], D(B) = [B^- ; B^+], A^+ = B^- \}$

Similarly,

DEFINITION 3 (continued):

$$[\![A \{r_1,\ldots,r_n\} B]\!] := \dot{\bigcup_{i=1}^{n}} [\![A \{r_i\} B]\!],$$

where the r_i are primitive relations.

Finally consider a time net for three or more A-intervals. We will define its set of models inductively assembling it from the models of certain subnets. We need some auxiliary terminology.

DEFINITION 4: Let $F = \{f_1, \ldots, f_n\}$, $f_i: X_i \to Y$, be a set of functions.

(i) F is *combinable*, iff $\forall\ 1 \leq i,j \leq n$: $\quad f_i|_{X_i \cap X_j} \equiv f_j|_{X_i \cap X_j}$.

(ii) If F is combinable, the function $\quad \text{Comb}(F): \bigcup_i X_i \to Y,$

defined by $\quad \text{Comb}(F)|_{X_i} \equiv f_i$, $i=1,\ldots,n$

is called the *combination* of F.

We then define

DEFINITION 3 (continued):

$$[\![\langle\{I_1,\ldots,I_n\}, C\rangle]\!] := \{\ \mathrm{Comb}(D, D_2, \ldots, D_n) \mid$$
$$D \in [\![\langle\{I_2,\ldots,I_n\}, C|_{\{I_2,\ldots,I_n\}^2}\rangle]\!],$$
$$D_2 \in [\![\langle\{I_1,I_2\}, C|_{\{I_1,I_2\}^2}\rangle]\!],$$
$$\ldots$$
$$D_n \in [\![\langle\{I_1,I_n\}, C|_{\{I_1,I_n\}^2}\rangle]\!],$$
$$\{D, D_2, \ldots, D_n\} \text{ is combinable } \}.$$

Notice that for each model of an arbitrary time net there is exactly one singleton time net of which it is also a model. The labels in the singleton time net are elements of the corresponding disjunctive relations in the original time net.

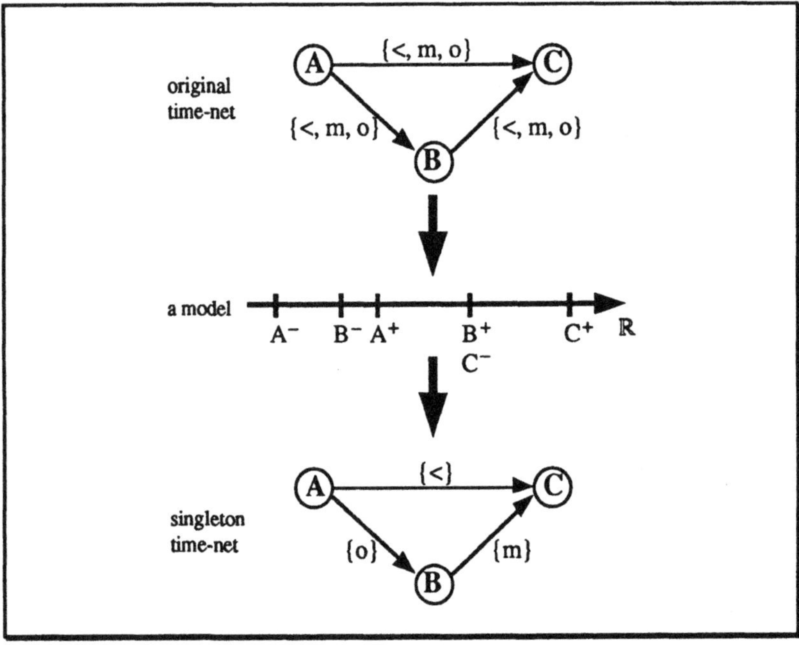

Fig. 21 – A time net and one of its singleton time nets

The first problem that we discuss is

PROBLEM 1 (Global consistency): Decide whether a given time net possesses a model. Or, equivalently: decide whether all intervals can be arranged on a linear time line in accordance with the (disjunctive) relations in the net.

Allen's and van Beek's path-consistency algorithms only check for consistency within all 3-cliques or 4-cliques of intervals (local consistency). [Allen83] contains an example time net that is globally inconsistent; the propagation algorithm, however, fails to detect the inconsistency, because it does not infer all consequences of the original labeling. The difference between local and global consistency is mirrored in the respective complexities: While Allen's incomplete propagation algorithm runs in $O(N^3)$ time, where N is the number of intervals, the following theorem holds:

THEOREM 1 [Vilain/Kautz86]: PROBLEM 1 is NP-complete.

Another problem related to satisfiability is

PROBLEM 2 (Minimal Labeling): For a given time net compute the minimal edge labeling, i.e. for each pair of intervals compute the disjunction of *exactly* those primitive relations that can participate in a singleton time net of a model of the net.

Again, it is clear that the normal path-consistency algorithm does not compute a minimal labeling. For any inconsistent time net the minimal labeling consists of empty disjunctions only. In the case of the counter-example from above, however, all edge labels remain non-empty after running Allen's algorithm. As in the case of PROBLEM 1 it is known that

THEOREM 2 [Vilain/Kautz86]: PROBLEM 2 is NP-complete.

Actually, PROBLEM 1 and PROBLEM 2 are polynomially transformable into each other.

Faced with theorems 1 and 2 there are basically three options, some of which may be available only under certain circumstances:

- *Do nothing*: accept the results of the path-consistency algorithms as an approximation. The price is loss of completeness.
- *Use exponential algorithms*: as [Valdes-Peres87] demonstrates it is not difficult to construct a backtracking algorithm which enumerates all singleton specializations of a given labeling and checks them for consistency. The price is loss of efficiency.
- *Restrict the language*: by making the language of interval relations smaller one can hope to eliminate the "pathological" cases which lead to incompleteness. The price is loss of expressive power.

Since none of the options is universally optimal, one can only make a pragmatic selection based on the properties of the problem at hand. Regrettably, to this date there exists little published material about the requirements of certain generic tasks or application scenarios on the temporal representation language. In their original paper [Vilain/Kautz86] the authors mention that in a planning context the fragment they study is not applicable, because it cuts away not only pathological but also critically necessary relations.

In this thesis we are concerned with representing dynamic behavior; since we want to perform extensive reasoning on TDS representations, the do-nothing option is not acceptable. Similarly, using an exponential algorithm qualifies only as a last resort, because it might severely limit the size of the examples tractable by our techniques. As a consequence, restricting the language seems promising provided that the expressiveness of the resulting language is still satisfactory.

Before we try to estimate which loss of expressive power we are willing to accept, we want to emphasize that by rights we can measure expressiveness only w.r.t. the full Allen relation algebra, *not* w.r.t. what we might want to be able to express ideally. To bring home this point we first take a look at some examples of what even the *full* algebra fails to capture.

3.2.4 Representing Dynamic Behavior - What Allen's Interval Relations Can't Do

Representing behavior that varies with time is a ubiquitous task in qualitative reasoning about dynamic systems. Probably the most common instance is qualitative simulation where a language is needed to represent the envisionment derived from the structural description of the device. In the literature several proposals for such a language can be found, e.g. the graph of process structures in QPT [Forbus84], or the state diagram in ENVISION [de Kleer/Brown84]. Alternatively, there are qualitative simulation systems (e.g. HIQUAL [Voß87]) which use the same history-oriented format that we developed for TDS in the last section[16]. Fig. 22 shows HIQUAL's output for a simple buzzer system.

---

[16] This is, incidentally, one of the reasons which make this approach so appealing for us.

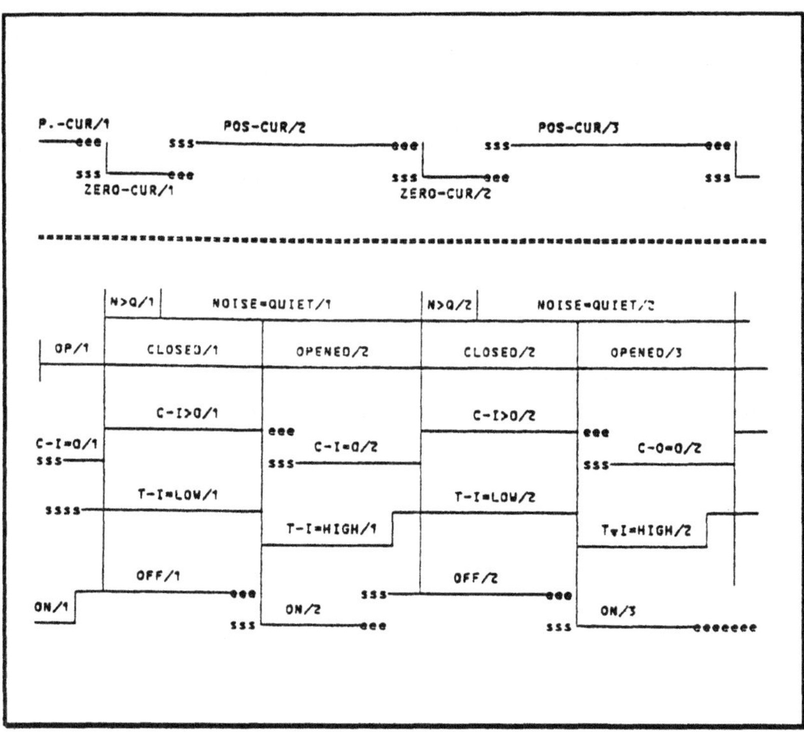

Fig. 22 – A section from HIQUAL's simulation output for a simple buzzer

Even in the extremely simple example of the buzzer the limitations of Allen relations become apparent. The fundamental insight about a buzzer that a qualitative simulator should discover is that its behavior repeats itself cyclically. This feature is obvious in a state diagram representation, but it is lost when histories are used. As we can see in fig. 22, all we can do is to transcribe an arbitrary finite path in the state diagram into the history format introducing new episodes for each time around the loop. Stated differently, there is no compact way of describing oscillating behavior in terms of Allen relations that avoids expanding the cycle[17].

Another example is shown in fig. 23:

[17] Actually, Allen relations do not show up all that badly in this example. We will return to fig. 22 in the next section where it serves to illustrate a particularly *good* feature of the interval algebra.

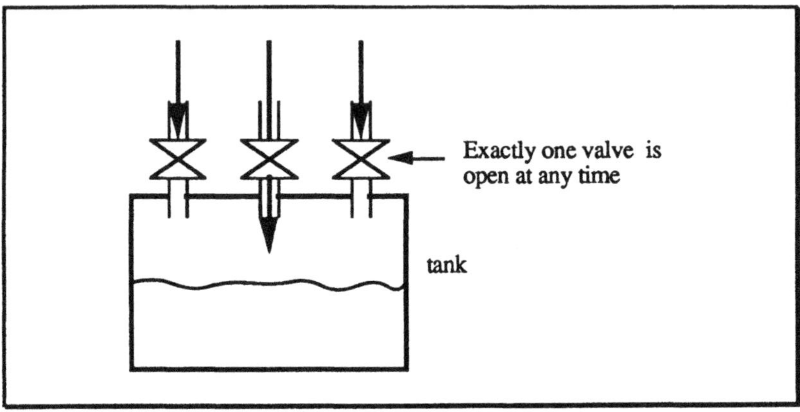

Fig. 23 – Problems with n-ary interval relations (n > 2)

Suppose that we associate a history with the state of each of the input valves of the tank and that we wish to express that by means of whatever mechanism exactly one valve must be open at any given time. For two valves we can state that

VALVE1-OPEN {m, mi} VALVE2-OPEN,

but for three or more episodes there is no analogous solution for specifying that all episodes must meet in any possible permutation without overlapping. This deficiency follows directly from the time net paradigm which allows statements about *pairs* of intervals to be made; the n-ary predicate needed in the example cannot be decomposed into a conjunction of binary relations.

Finally, consider a case (fig. 24) where during Allen-based episode propagation the result cannot be expressed in the history format, although the inputs can; the behavioral constraint is just a logical OR.

Here we encounter the problem that the output history may or may not contain an episode with value 0 depending on whether A's first and B's last episode meet or not. The basic assumption in Allen's temporal logic is that the presence of the events themselves (or rather, their corresponding intervals) is not subject to uncertainty; only the exact temporal order may not be known in full detail. There is thus no provision for optional episodes in the language.

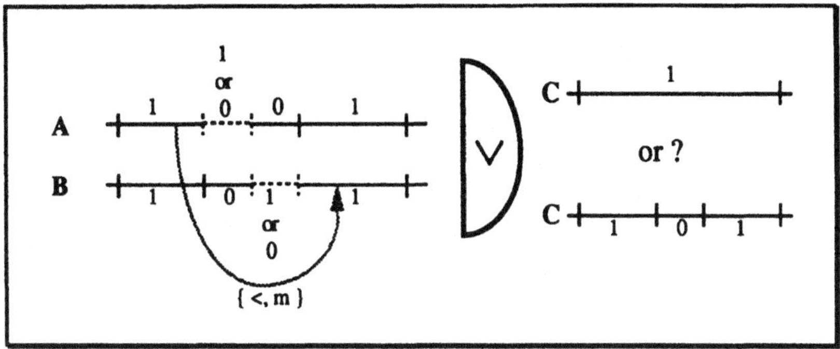

Fig. 24 – Problems with optional episodes

In summary, we can say that even the full interval relation algebra has a number of limitations which are, of course, inherited by every subset. In the following discussion of convex relations we are therefore concerned only with *further* losses of expressive power.

3.2.5 Convex Relations

Even in their limited form disjunctive interval relations are an essential feature of Allen's calculus. In TDS descriptions they can be viewed as a simple abstraction mechanism that helps to neglect insignificant constraints on the temporal order of the constituent episodes. In the output of qualitative simulation programs disjunctions appear as compact representations for *sets* of paths through the envisionment. As an example fig. 22 contains several occurrences of "sss" and "eee" indicating episodes whose starting point or end point may range across a set of positions according to different interval relations in the simulation result.

In theory any of the 2^{13} disjunctive interval relations would be expected to occur in a model specification or in the output of a qualitative simulator. The same is true for TDS descriptions and other representations of dynamic behavior. Contrary to this expectation an empirical examination of models, simulation results and TDS descriptions has revealed a much more lop-sided distribution: in practice only relatively few disjunctions ever play a role. In his thesis Hans Voß [Voß87] presents a possible explanation for this observation: the main use of disjunctions in model specifications is to give a temporal interpretation to causality. Conservatively speaking, "causes(Event1, Event2)" implies that the episode interval of the effect (Event2) cannot start before the beginning of the episode interval of the cause (Event1). Encoding this in Allen's interval relations we obtain

$$I_{Event2} \{=, s, si, d, oi, f, mi, >\} I_{Event1}.$$

In Voß' terminology this disjunction is called "not-starts-before". Similar arguments (e.g. about the temporal relationship between an event and its subevents) lead to other disjunctions which are also frequently found in examples, e.g.

{=, s, si}	"starts simultaneously"
{=, s, si, d, oi, f}	"starts within"
...	

Our next goal is to demonstrate that all these "typical" disjunctions share a characteristic so that we can later prove a favorable complexity result about them. We will approximate the definition of this common property in two steps, starting with an intuition-based formulation, which turns out to be too weak, and then refining it appropriately.

Notice that the three relations have the following convexity property in common: any two models[18] of a time net "A r B" where r is one of these relations can be transformed into one another by "continuously deforming" the R-intervals such that all intermediate stages of the transformation are also models of the time net.

Let us illustrate this property with a simpler example: suppose that the A-intervals A and B are specified to stand in one of the relations "before", "meets", and "overlaps". Consider those models from $[\![A \{<, m, o\} B]\!]$ in which B^-, B^+, and A^- are fixed at some arbitrary reals, i.e. the models differ only in the position they assign to A^+. Fig. 25 shows that the set of positions for A^+ in different models is convex in the normal set-theoretic sense, regardless of the particular choice for the other three end points. We call this property of interval relations *1-point-convexity*.

If on the other hand we restrict the relation between A and B to {<, o} we exclude all models that have A's right and B's left end points coincide. It follows that a model in which A's right end point comes before B's left end point cannot be transformed continuously into one in which it comes after it, because there is an intermediate stage which is not a model of A {<, o} B. Hence 1-point-convexity is violated for {<, o}.

---

[18] Recall definition 3 in section 3.2.3.

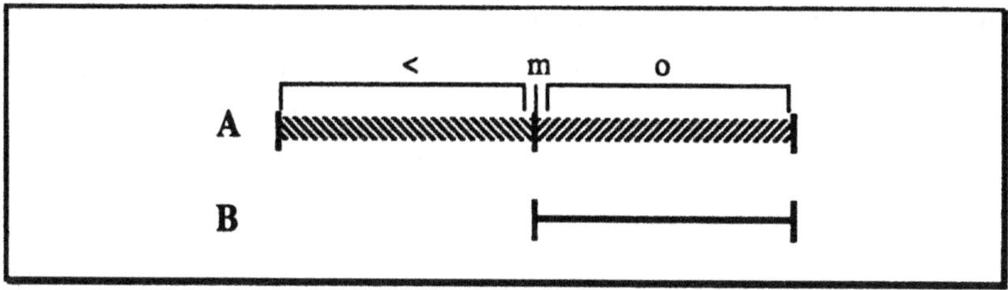

Fig. 25 - Possible models of A {<, m, o} B (shaded areas indicate possible positions for A^+)

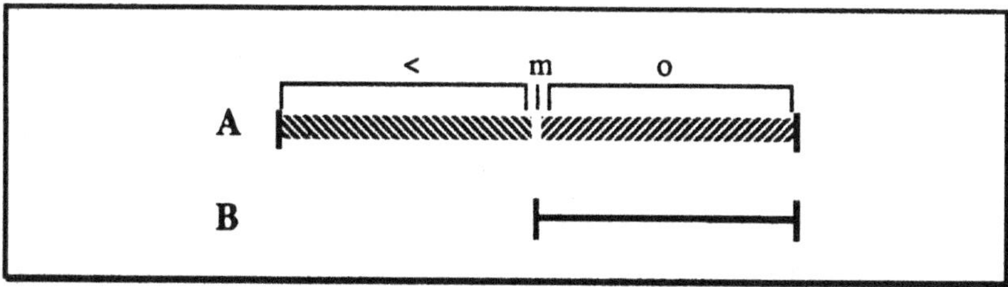

Fig. 26 - Possible models of A {<, o} B

The next example shows that 1-point-convexity does not yet capture the stronger, intended notion. Take the time net "A {<, m, mi, >} B": clearly, not all models can be transformed continuously into each other, yet this is true within each class of models in which three end points are fixed. However, the property we have in mind calls for *all* models to be continuously transformable into each other, not just some.

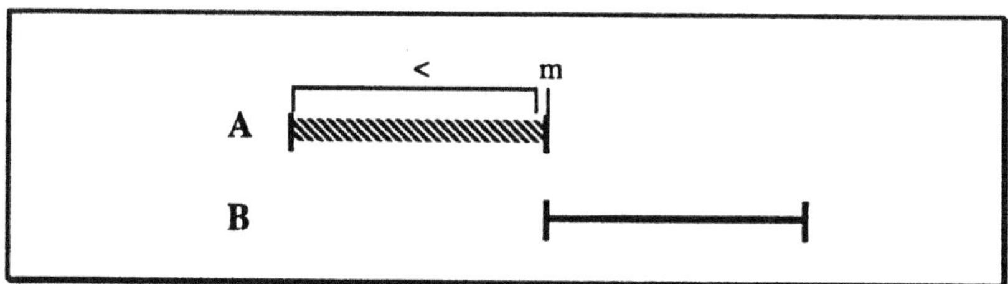

Fig. 27 - Possible models of A {<, m, mi, >} B. Given the initial choice of A^-, B^- and B^+, models for mi and > are not generated.

The problem arises because in the case of "A {<, m, o} B" we vary across all models (modulo order-preserving shifts of the end points) even though we keep three of the end points fixed. In fig. 27 we don't. In order to generate *all* models of "A {<, m, mi,

>} B" at least *two* end points have to change places in the order. Hence we have to define how two models are continuously transformed into each other keeping the R-interval of one A-interval fixed while moving *both* end points of the other. Before we give a formal definition let us again look at a graphic example. This time we visualize the models of a time net in a two-dimensional diagram adopted from a similar representation in [Rit88].

We consider time nets "A r B" where r is a primitive Allen relation and consider only those models in which A is mapped to a fixed interval $[A_0^- ; A_0^+]$. For each primitive r we define the set

$$B(r,A_0) := \{ D(B) \mid D \in [\![A\, r\, B]\!], D(A) = [A_0^- ; A_0^+] \} \subseteq \text{Intervals}(\mathbb{R}).$$

In fig. 28 each element $[B_0^- ; B_0^+] \in B(r,A_0)$ corresponds to the point at the intersection of B_0^- and B_0^+ on the two real time axes. Because always $B_0^- < B_0^+$, all models are mapped to points in the upper left half plane. Furthermore we see that for each primitive relation r the set $B(r,A_0)$ is represented by a zero-, one-, or two-dimensional region. Not surprisingly the sets of models for time nets with a disjunctive relation are represented by the union of the regions corresponding to the individual primitive relations.

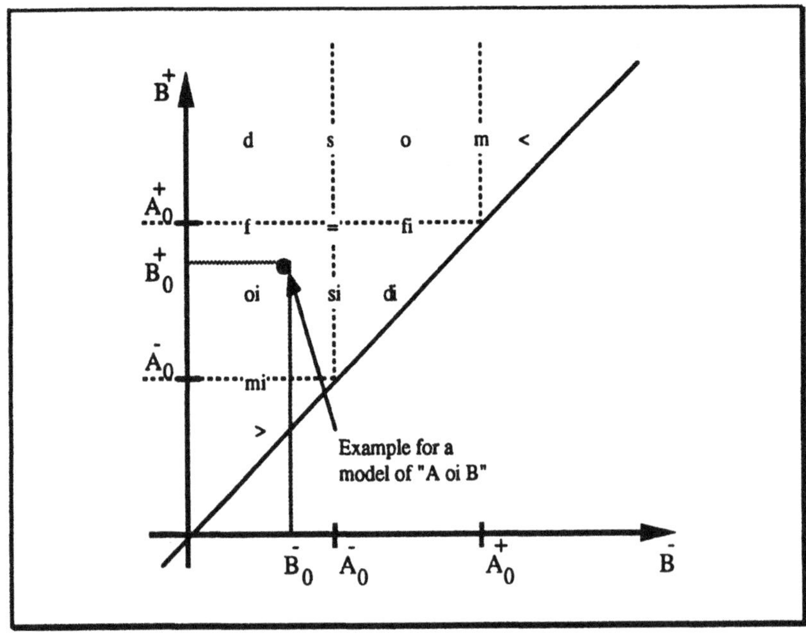

Fig. 28 - Graphical representation of $B(r,A_0)$ for different relations r.

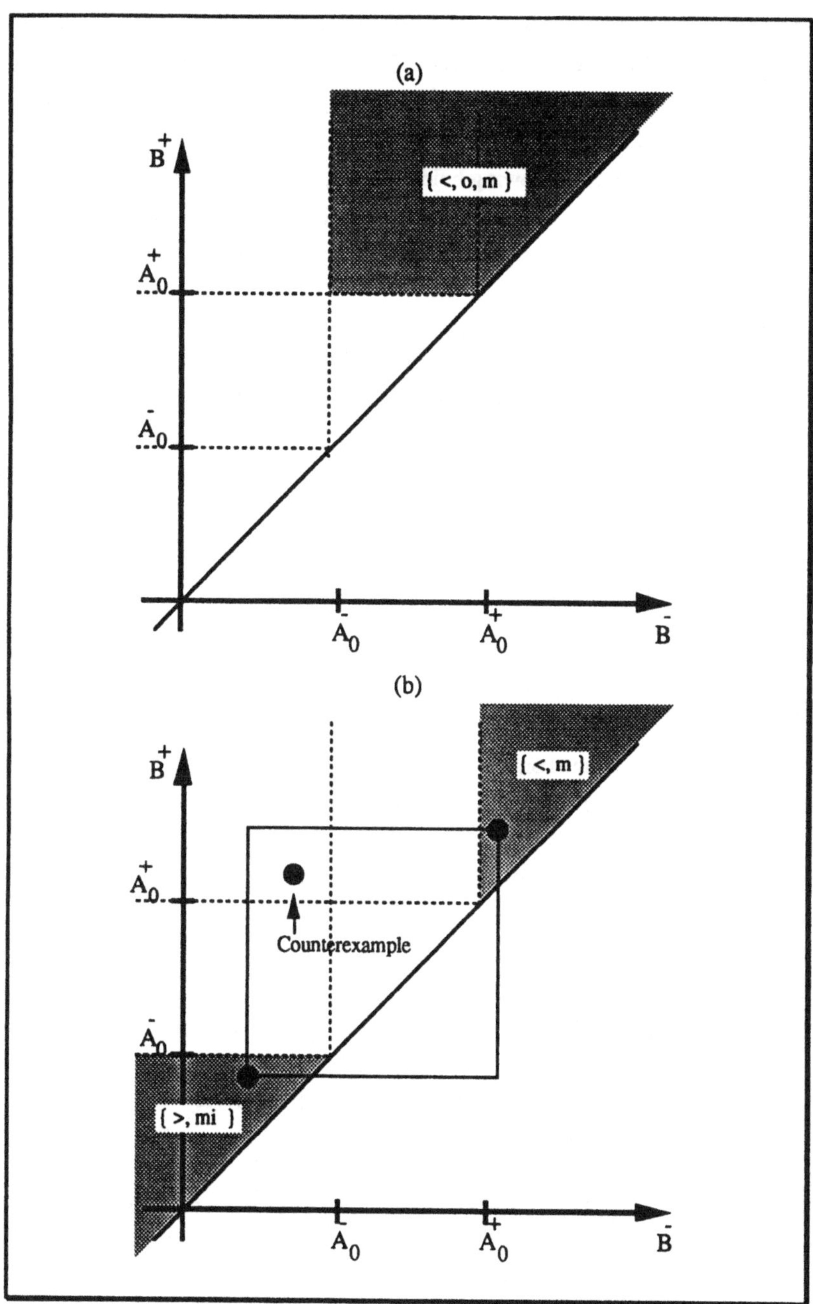

Fig. 29 - Applying the 2-point-convexity criterion to (a) A {<, o, m} B and (b) A {<, m, mi, >} B.

Based on the diagram we can generalize the notion of 1-point-convexity. Let $\circ$ be a disjunctive interval relation and $D_1, D_2 \in [\![A \circ B]\!]$ s.t. $D_1(A)=D_2(A)=[A_0^- ; A_0^+]$, $D_1(B)=[B_1^- ; B_1^+]$ and $D_2(B)=[B_2^- ; B_2^+]$. We say that D_1 can be continuously transformed into D_2, iff for each point $\langle B_3^- ; B_3^+\rangle$ in the rectangle defined by $\langle B_1^- ; B_1^+\rangle$ and $\langle B_2^- ; B_2^+\rangle$ in fig. 28 the following holds: if $[B_3^- ; B_3^+]$ belongs to *any* $B(r,A_0)$ (i.e. $\langle B_3^- ; B_3^+\rangle$ lies in the upper left half plane), then D_3 defined by $D_3(A)=[A_0^- ; A_0^+]$ and $D_3(B)=[B_3^- ; B_3^+]\rangle$ is also a member of $[\![A \circ B]\!]$. We call this stronger property *2-point-convexity*.

Fig. 29 shows that 2-point-convexity is satisfied by A {<, m, o} B, but not by A {<, m, mi, >} B. Indeed, all examples that we studied fall into two classes. In the larger class all relations are 2-point-convex. In the other class the behavior cannot be formalized adequately even when we use the full set of Allen relations (e.g. the tank example in section 3.2.4). It seems safe to say then that the set of 2-point-convex relations is sufficient to represent a large proportion of the behaviors which are not inherently beyond the expressive power of Allen relations.

From now on we will be concerned with 2-point-convex relations exclusively. We will call them simply *convex relations*.

We shall now investigate the computational properties of convex relations. After all, the motive behind the whole enterprise is to find a subset of the complete Allen algebra that is *both* sufficiently expressive *and* computationally tractable.We begin by giving a formal reconstruction of 2-point-convexity so that we can prove results about it. In section 3.2.3 we already defined models for time nets (definition 3). We first define two operations on models: projection and restriction.

DEFINITION 5: Let M be a set of models for the time net $\langle V, C\rangle$. Let $I \in V$.

Then the set

$$\mathrm{Proj}(M,I) := \{ \langle I^-, I^+\rangle \mid D \in M, D(I) = [I^- ; I^+] \} \subseteq \mathbb{R} \times \mathbb{R}$$

is called the *projection* of M onto I.

DEFINITION 6: Let M be a set of models for the time net $\langle V, C\rangle$. Let $I \in V$, $c_1, c_2 \in \mathbb{R}$.

Then the set

$$\mathrm{Rest}(M,I,c_1,c_2) := \{ D \in M \mid D(I) = [c_1;c_2] \}$$

is called the*restriction* of M to $I \mapsto [c_1; c_2]$.

DEFINITION 7: A set $M \subseteq \mathbb{R} \times \mathbb{R}$ is *interval-convex* iff

$\forall (x_1,y_1), (x_2,y_2) \in M : \forall x, y \in \mathbb{R}:$

$[(x_1 \le x \le x_2 \vee x_2 \le x \le x_1) \wedge (y_1 \le y \le y_2 \vee y_2 \le y \le y_1) \wedge (x < y)$

$\Rightarrow (x,y) \in M].$

LEMMA 1 (without proof): Let $M \subseteq \mathbb{R} \times \mathbb{R}$.

M interval-convex

$\Leftrightarrow \exists M_1, M_2 \subseteq \mathbb{R}$: M_i convex $\wedge\ M = (M_1 \times M_2) \cap \{(x,y) \mid x,y \in \mathbb{R}, x < y\}$.

Now we can define convex interval relations.

DEFINITION 8: A disjunction $\circ$ of primitive Allen relations is *convex*, iff

$\forall$ A,B: $\exists c_1, c_2 \in \mathbb{R}$: $c_1 < c_2 \wedge$ Proj(Rest($[\![A \circ B]\!]$,A,c_1, c_2),B) is interval-convex.

Note that there is nothing special about c_1 and c_2. If the condition holds for one pair c_1, c_2 then the same follows for any other pair $c_1 < c_2$. This means that the $\exists$-quantifier could be replaced equivalently by a $\forall$-quantifier. The seeming asymmetry of A and B in definition 8 is resolved in the corollary to theorem 5.

Although earlier in our discussion 1-point-convexity was seen to be insufficient for the characterization of convex relations, interestingly it can still be used to give a more graphic account of convex interval relations: We can identify pairs of primitive Allen relations which are immediate neighbors in this sense and represent the result in the graph shown in fig. 30 where relations connected by an edge can be transformed into each other directly (i.e. without going via a third relation). If we reinterpret definition 8 in the graph, we get: $\circ$ is convex iff for any primitive $r_1, r_2 \in \circ$ all relations along every shortest path from r_1 to r_2 are also members of $\circ$.[19]

Alternatively, we can view the graph as the Hasse diagram of an ordering $\sqsubseteq$. Due to the symmetry of the graph, the convex interval relations are the "intervals" w.r.t. $\sqsubseteq$, and we have: $\circ$ is convex $\Leftrightarrow \exists r_1, r_2$: $\circ = \{ r \mid r_1 \sqsubseteq r \sqsubseteq r_2 \}$.

By counting these "intervals" we find that there are 82 convex relations.

---

[19] The correctness of this claim can easily be checked using the equivalence result of theorem 5.

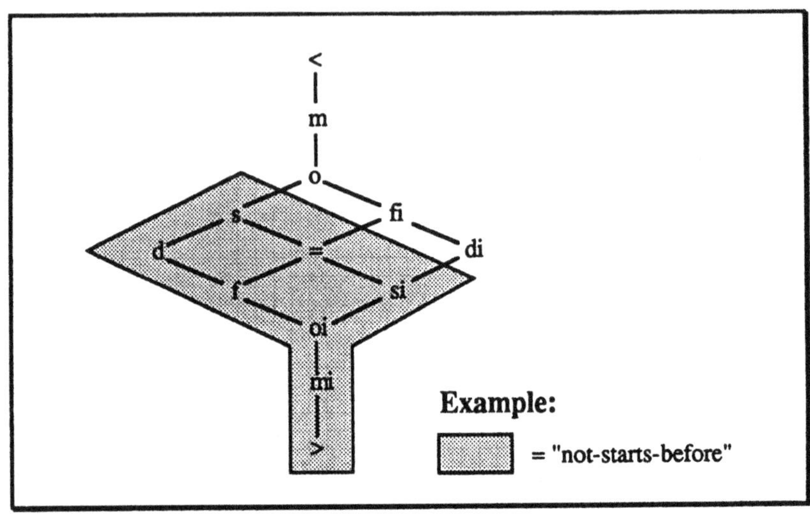

Fig. 30 - The continuous transformability relation on primitive Allen relations

We will now prove that in the set of convex relations the problems PROBLEM 1 (global consistency) and PROBLEM 2 (minimal labeling) posed in section 3.2.3 can be solved in polynomial time. The plan is to prove that convex relations are identical to a subset of Allen relations that has been characterized before in the literature and for which the desired results are known. Although these results are in itself of theoretical interest, they gain practical value only, if one can find a characterization of the subset in terms of an application where they can be used. Our goal in the study of convex relations is to provide such a characterization.

To provide the basis for the equivalence proof we briefly review this alternative characterization and the complexity results pertaining to it.

When we defined the semantics of Allen relations we made use of the fact that each primitive relation can be specified as a conjunction of certain end point orderings. Similarly a disjunctive Allen relation can be stated as a disjunction of conjunctions of end point orderings. We can extend the construction to the syntactic level and talk about time nets in which the nodes are symbolic time *points* and the edges are labelled with disjunctions of the primitive time *point* relations <, = and >. The same questions as for interval-based time nets can also be asked for point-based time nets. Even the same algorithms (e.g. Allen's propagation algorithm) can be used. However, the complexity results are different.

In their original paper Vilain and Kautz proved

THEOREM 3 [Vilain/Kautz86]: In the time point algebra PROBLEM 1 can be solved in polynomial time (using Allen's algorithm).

For some time it was thought that the same was true for PROBLEM 2. Surprisingly van Beek came up with a counter example [van Beek89] demonstrating that Allen's algorithm is still incomplete for the full time point algebra. On the other hand he was able to show that there is a subset of the time point algebra – called the continuous point relations – where Allen's algorithm does compute the minimal labeling. The result is:

THEOREM 4 [Vilain et al. 90]: In the set of continuous point relations PROBLEM 2 can be solved in polynomial time (using Allen's algorithm).

The set of continuous point relations differs from the full point algebra in that only the point relations {<}, {=}, {>}, {<, =}, {>, =}, and {<, =, >} are allowed, but not {<, >}.

Both theorems can be carried over to interval relations in the sense that both the full time point algebra and the set of continuous point relations can be translated equivalently to some subsets of the interval relation algebra and vice versa. The translation is defined as follows. Let

$$P_1 := \{ \{<\}, \{=\}, \{>\}, \{<, =\}, \{>, =\}, \{<, =, >\} \} \qquad \text{and}$$
$$P_2 := P_1 \cup \{ \{<, >\} \}.$$

We associate with every A-interval I two symbolic time points (its starting and end points) which – abusing notation – we denote by the symbols I^- and I^+, too.

DEFINITION 9: Let P be a set of time point relations.

An interval relation $\circ$ is *P-definable* iff:

$$\begin{aligned} A \circ B \Leftrightarrow\quad & A^- r_1 B^- \wedge \\ & A^- r_2 B^+ \wedge \\ & A^+ r_3 B^- \wedge \\ & A^+ r_4 B^+, \text{ for some } r_i \in P \ (i=1,\ldots,4). \end{aligned}$$ [20]

---

20 The relations $A^- < A^+$ and $B^- < B^+$ are part of the axiomatization of intervals and are therefore not duplicated here.

Using this translation we can trivially state the following theorems:

THEOREM 3': In the set of P_2-definable interval relations PROBLEM 1 can be solved in polynomial time (using Allen's algorithm).

THEOREM 4': In the set of P_1-definable interval relations PROBLEM 2 can be solved in polynomial time (using Allen's algorithm).

Our main result is the following equivalence:

THEOREM 5: Let $\circ$ be an interval relation. $\circ$ is convex $\Leftrightarrow$ $\circ$ is P_1-definable.

PROOF: "$\Rightarrow$": Assume that $\circ$ is convex. We construct an equivalent relation on interval end points using only relations from P_1. Let A and B be A-intervals and consider the time net "A $\circ$ B". Choose $c_1, c_2 \in \mathbb{R}$, $c_1 < c_2$, arbitrarily (the construction is independent of the particular choice) and define

$$M := \mathrm{Proj}(\mathrm{Rest}([\![A \circ B]\!], A, c_1, c_2), B).$$

From definitions 7 and 8 and lemma 1 it follows that

$$M = (M_1 \times M_2) \cap \{(x,y) \mid x,y \in \mathbb{R}, x < y\}, \text{ for some convex } M_i \subseteq \mathbb{R}.$$

The M_i possess a natural interpretation in fig. 28. Since $\circ$ is an interval relation, M must be the union of some of the regions marked in the diagram. Then M_1 is the projection of M onto the B^--axis and M_2 its projection onto the B^+-axis.

Consider M_1 first.

Case I (M_1 singleton):

From fig. 28 we see that the only possible projections of M onto the B^--axis, which are singletons, are $\{c_1\}$ and $\{c_2\}$.

If $M_1=\{c_1\}$, add "$B^- \{=\} A^- \wedge B^- \{<\} A^+$" to the end point relation.

If $M_1=\{c_2\}$, add "$B^- \{=\} A^+ \wedge B^- \{>\} A^-$".

Case II (M_1 non-singleton interval):

Again we see from fig. 28 that c_1 and c_2 are the only possible end points of M_1, in addition the interval can be unbounded on cne or both sides. Each of these cases can be translated into literals of the end point relation:

M_1	**Translation to end point relation**
$(-\infty, c_1)$	$B^- \{<\} A^- \wedge B^- \{<\} A^+$
$(-\infty, c_2)$	$B^- \{<, =, >\} A^- \wedge B^- \{<\} A^+$
$(-\infty, \infty)$	$B^- \{<, =, >\} A^- \wedge B^- \{<, =, >\} A^+$
(c_1, c_2)	$B^- \{>\} A^- \wedge B^- \{<\} A^+$
(c_1, ∞)	$B^- \{>\} A^- \wedge B^- \{<, =, >\} A^+$
(c_2, ∞)	$B^- \{>\} A^- \wedge B^- \{>\} A^+$

$(-\infty, c_1]$	$B^- \{<\} A^- \wedge B^- \{<, =\} A^+$
$(-\infty, c_2]$	$B^- \{<, =, >\} A^- \wedge B^- \{<, =\} A^+$
$[c_1, c_2)$	$B^- \{>, =\} A^- \wedge B^- \{<\} A^+$
$(c_1, c_2]$	$B^- \{>\} A^- \wedge B^- \{<, =\} A^+$
$[c_1, c_2]$	$B^- \{>, =\} A^- \wedge B^- \{<, =\} A^+$
$[c_1, \infty)$	$B^- \{>, =\} A^- \wedge B^- \{<, =, >\} A^+$
$[c_2, \infty)$	$B^- \{>\} A^- \wedge B^- \{>, =\} A^+$

M_2 is treated in the same way substituting B^+ for B^- in cases I and II. The translations of M_1 and M_2 combine to form a complete specification of an end point relation equivalent to $\circ$.

"$\Leftarrow$": Assume that $\circ$ is P_1-definable. Choose $c_1, c_2 \in \mathbb{R}$, $c_1 < c_2$, arbitrarily. From the equivalent end point relation it follows that $A \circ B \Rightarrow B^- r_1 c_1 \wedge B^- r_2 c_2$. Let M_1 be the set of B^- which satisfy this condition (analogously for B^+ and M_2). The sets$\{x \in \mathbb{R} \mid x\, r\, c\}$ are convex for arbitrary $r \in P_1$, $c \in \mathbb{R}$ and this holds for their intersection, too, of course. Combining the results for M_1 and M_2 we get $\text{Proj}(\text{Rest}([\![A \circ B]\!], A, c_1, c_2), B) = (M_1 \times M_2) \cap \{(x,y) \mid x,y \in \mathbb{R}, x < y\}$, which implies the conjecture. ■

COROLLARY: Using the proof in the "$\Leftarrow$"-direction it can be shown that for convex $\circ$ $\text{Proj}(\text{Rest}([\![A \circ B]\!], B, c_1, c_2), A)$ is interval-convex, too. Hence, either of the two dual conditions can be employed in definition 8.

Summarizing our results we can say that convex relations are a subset of the full Allen interval relation algebra which is sufficiently expressive for our purposes and for which Allen's propagation algorithm solves the global consistency and minimal labeling problems in polynomial time.

3.2.6 Putting Everything Together

Now that we have identified a practical way of representing temporal constraints in a quantity-oriented TDS description we can complete our definition.

DEFINITION 10: A *TDS description* is a triple $\langle Q, H, C \rangle$ where

- Q is a finite set of quantities;
- $H = \{H_q \mid q \in Q\}$ is a set of value histories and
- $C = \{C_{E,E'} \mid E, E'$ episodes of histories in $H\}$ is a set of temporal constraints. Each $C_{E,E'}$ is a convex relation specifying the primitive Allen relations allowed between the intervals of E and E'.

DEFINITION 11: For each TDS description $S = \langle Q, H, C\rangle$ there is an associated *TDS time net* $N(S) = \langle V, C\rangle$ defined by

$$V := \{ I \mid \langle I, v\rangle \in H_q \in H \}$$

(C is used directly as the labeling function).

The translation of our standard example is shown in fig. 31. The TDS description consists of two histories - one for each quantity - and the set of temporal constraints which form the TDS time net. Some of the temporal constraints are shown as dotted lines; the complete labeling function is given as fig. 32. The temporal constraints fall into two categories. The episodes of the same history are linked by chains of "meets"-relations. These relations merely enforce that histories do not have gaps. The other group of relations are those connecting episodes of *different* histories; they correspond to the causal relationships that we want to capture by the TDS and constrain the relative position of the two histories w.r.t. each other.

Notice again that the relation between the initial episodes of the two histories are constrained to {o, s, d} although from figure 11 it would appear to be {o}. This is, however, an artifact of the diagram: the salient features of the TDS are the 1-0- and 0-1-transitions, not the positions of the starting (or end) points of the histories. In general it

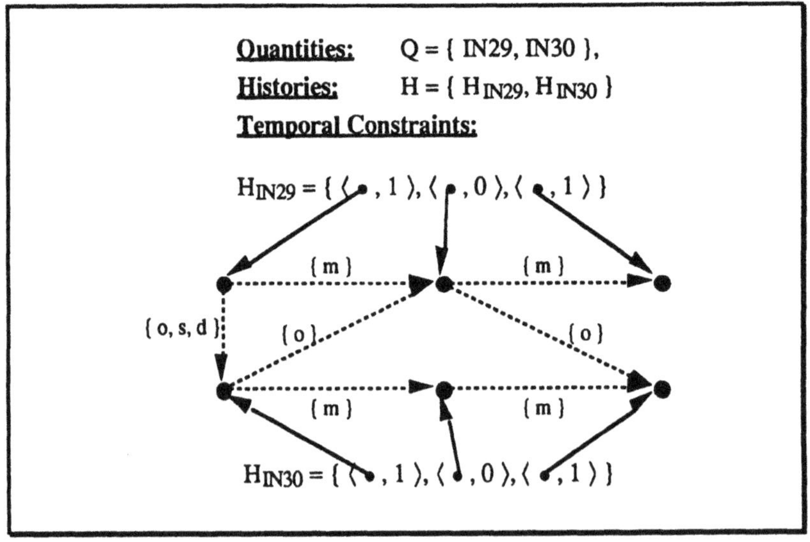

Fig. 31 - The representation of fig. 11 as a TDS description

is not meaningful to specify an order on the starting and end points of histories[21] and – instead of formally forbidding it – we assume that it is not done.

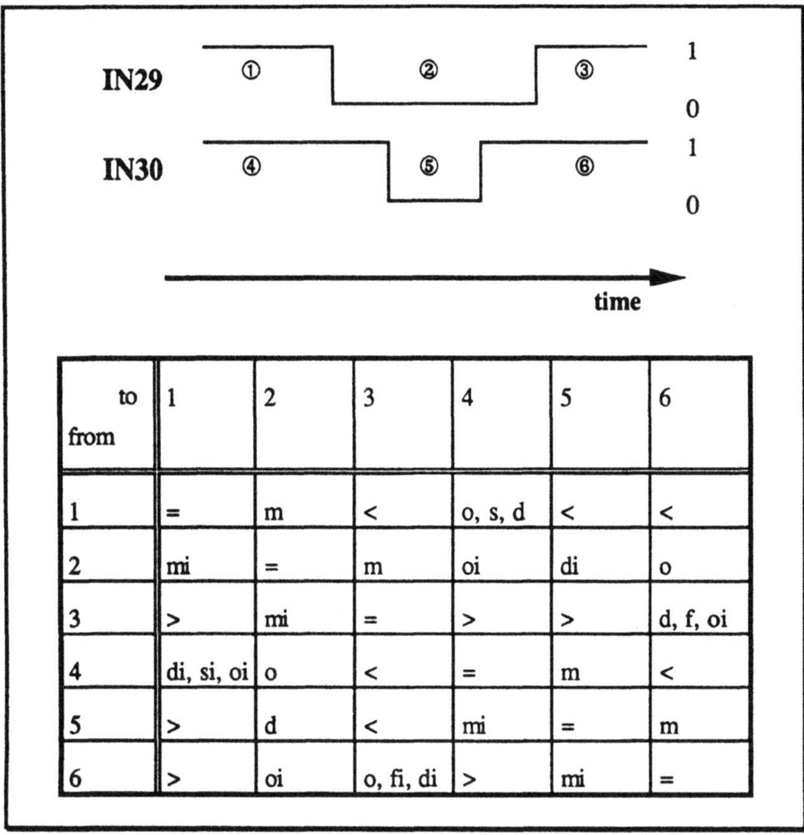

to from	1	2	3	4	5	6
1	=	m	<	o, s, d	<	<
2	mi	=	m	oi	di	o
3	>	mi	=	>	>	d, f, oi
4	di, si, oi	o	<	=	m	<
5	>	d	<	mi	=	m
6	>	oi	o, fi, di	>	mi	=

Fig. 32 – The complete TDS time net for the IN29/IN30 example in tabular form. The diagram introduces episode numbers for easier reference

We will extend definitions 10 and 11 in section 3.7 where we deal with quantitative information (bounds on episode durations). Notice further that the definitions can easily be generalized to episodes with more than one value; these episodes are convenient as a compact representation of alternative behaviors or as a specification of don't care-episodes in the middle of a history.

---

[21] If one needed to do so, one would also have to include the episodes before (after) in the history, and then the point in question would no longer be the starting (end) point of the history.

3.3 TDS Instances and Reality

Put in event-recognition terminology TDS descriptions are the patterns against which the observations are matched. The goal is to accumulate evidence for an occurrence of the given TDS. The next step on the way to a definition of temporal matching will therefore be a formalization of "the REAL flow of events" and particularly of TDS occurrences.

As before we assume that time is densely ordered without least or greatest elements, e.g. isomorphic to $\mathbb{R}$. Our view of "reality" is that of a universe of quantities U_Q together with an assignment function VALUE that maps from quantities and time points to values:

$$\text{VALUE} : U_Q \times \mathbb{R} \rightarrow \dot{\bigcup_{q \in U_Q}} \text{dom}(q)$$

Just as in the definition 10 we require that quantities change their values only finitely often in any bounded interval, i.e. for each $q \in U_Q$ we can partition $\mathbb{R}$ into a set of non-overlapping, contiguous intervals during each of which the function VALUE(q,.) is constant. We call these intervals *reality-episodes*.

Although VALUE is a total function in both its arguments, in practice it is known only for certain discrete values of its arguments: if a quantity is observable and we decide to measure it at a particular time point, then we observe the value assigned to this combination by VALUE.[22] It may seem unintuitive to say that VALUE is total, because $\mathbb{R}$ is unbounded. As it is, we never make any statements about VALUE at time points in the future so that we can simply define VALUE as the limit of a series of partial functions which become gradually more defined as history unfolds. Even if we admit branching future assignments as approximations (in the sense of [McDermott82]), in the limit we are left with a single non-branching path that corresponds to "the way things have turned out to be like". Either way VALUE is well-defined.

Each TDS description can be viewed as a generic pattern for infinitely many instances depending on which model of the temporal constraints one chooses.

---

[22] The effects of noisy data and similar sources of uncertainty are not discussed in this thesis.

DEFINITION 12: Let S=⟨Q, H, C⟩ be a TDS description and N(S)=⟨V, C⟩ its associated TDS time net.

(i) A triple TI=⟨Q, H, D⟩ is called an *instance* of S iff $D \in [\![N(S)]\!]$.

(ii) The interval $[t_{min} ; t_{max}]$

$$\text{where} \quad t_{min} := \min_{I \in V} I^- \qquad \text{and} \qquad t_{max} := \max_{I \in V} I^+$$

is called the *basis interval* BASIS(TI).

An example is given in fig. 33.

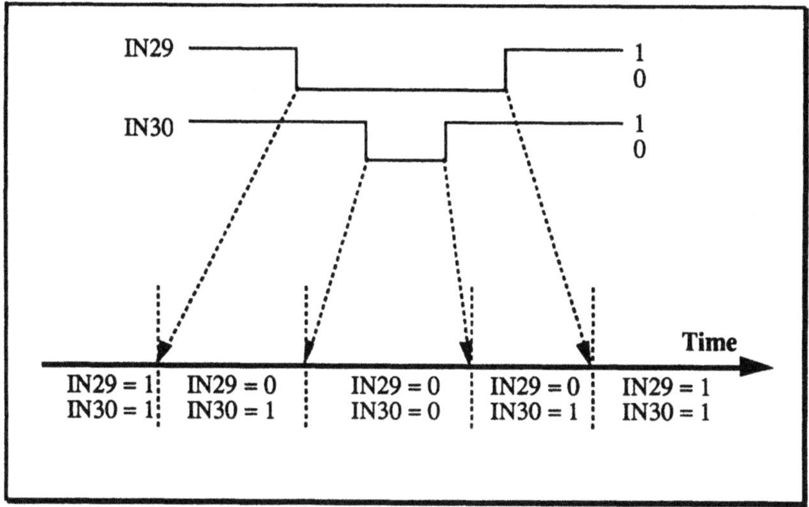

Fig. 33 - The relation between a TDS description and one of its instances

DEFINITION 13: An instance ⟨Q, H, D⟩ of the TDS description ⟨Q, H, C⟩ *occurs* in the interval $O \subset \mathbb{R}$, iff

(i) $O = \text{BASIS}(\langle Q, H, D\rangle)$ and

(ii) $\forall t \in O, q \in Q$: $(\text{VALUE}(q,t) = v \Rightarrow \exists I: \langle I,v\rangle \in H_q \wedge t \in [I^- ; I^+])$.

We omit the interval when we are not interested in the particular instance that is occurring.

3.4 Measurements

Compared to TDSs the formalization of the data (measurements) is almost trivial.

By measurement we mean the *result* of performing a measuring action, not the act itself. We use the term in a broader sense than usual: although most of the quantities we are concerned with are *observable* quantities (such as IN29 and IN30 in our running example), others are *adjustable*.[23] These quantities play an important role as inputs in experiments where characteristic changes in the measurable quantities are induced by setting adjustable quantities to certain values. An example is given in the introduction. It turns out that from the perspective of representation both kinds of quantities can be treated in the same way by a slight generalization of the meaning of measurements: measurements are simply data about the value of a quantity at a given time point, regardless of whether they are the result of a measuring action or whether they have been brought about by setting the quantity to that value. Apart from the fact that according to the type of quantity the measurement suggestions produced by the temporal matching algorithm have to be verbalized differently, the differentiation has no consequences.

DEFINITION 14: A *measurement* is a triple $\langle q, t, v\rangle$, where $q \in U_Q$, $t \in \mathbb{R}$ and $v = \text{VALUE}(q,t) \in \text{Dom}(q)$.

DEFINITION 15: A *measurement sequence* M is a finite set of measurements $\{\langle q_i, t_i, v_i\rangle\}_{i=1,\ldots,n}$ where $t_1 < t_2 < \ldots < t_n$.
Let *Int(M)* $:= [t_1 ; t_n] \subseteq \mathbb{R}$.

DEFINITION 16: A measurement sequence $M = \{\langle q_i,t_i,v_i\rangle\}_{i=1,\ldots,n}$ is *compatible* with an instance $\langle Q, H, D\rangle$ of the TDS description S, iff
$\forall i: \; (t_i \in \text{BASIS}(\langle Q, H, D\rangle) \;\Rightarrow\; \exists \langle I, v_i\rangle \in H_{q_i}: \; t_i \in [I^- ; I^+])$.

Throughout the text we will use the terms "measurement" and "observation" synonymously.

---

[23] Still other parameters of a technical device will be neither observable nor (directly) adjustable, but it would make little sense to include any of these in a TDS description.

3.5 Matching - A Definition

In the previous section we have formalized the environment in which we want to detect occurrences of TDSs so that we can now rephrase the task more formally. We will do this in two steps:

1) In this section we assume that we have already collected a measurement sequence and define under what circumstance it can serve as evidence for or against a given TDS.

2) In the next section we will take the a posteriori definition and derive from it an algorithm which plans measurements and matches observations against expectations in such a way that we will eventually arrive at a measurement sequence which satisfies the definition provided that we do not detect a discrepancy earlier in the matching process.

What we hope to define eventually is a predicate "matches" that holds between measurement sequences and TDS descriptions and captures the way in which observations can help rule out occurrences of certain TDS descriptions and support others. Ideally, we would like to identify exactly those pairs of measurement sequences M and TDS descriptions S where the observation of M implies that an instance of S is occurring. In other words: if we call this proposition "M determines S", then we would like to define "M matches S" in such a way that two properties hold:

Completeness:

$\forall S\ \forall M$: M determines S $\Rightarrow$ M matches S.

Soundness:

$\forall S\ \forall M$: M matches S $\Rightarrow$ M determines S.

Unfortunately, this is impossible, as a simple example shows. As long as we are committed to sequences of discrete measurements only, we can never say with certainty what has happened between the individual measurements. Consider our example TDS and the measurement sequence in fig. 34.

Clearly, the observations are compatible with the TDS description so that we cannot rule it out. Neither can we guarantee that an instance of this particular TDS description must have occurred. Indeed, M is equally compatible with instances of other TDS descriptions as fig. 35 shows.

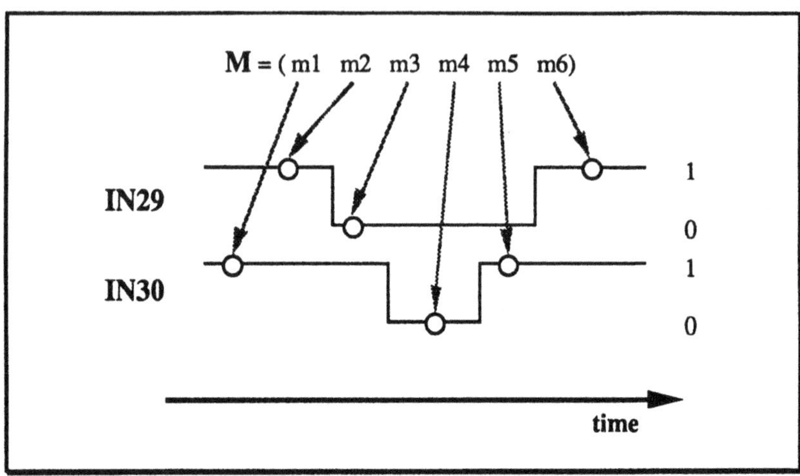

Fig. 34 – A sparse measurement sequence M which is compatible with an instance of S

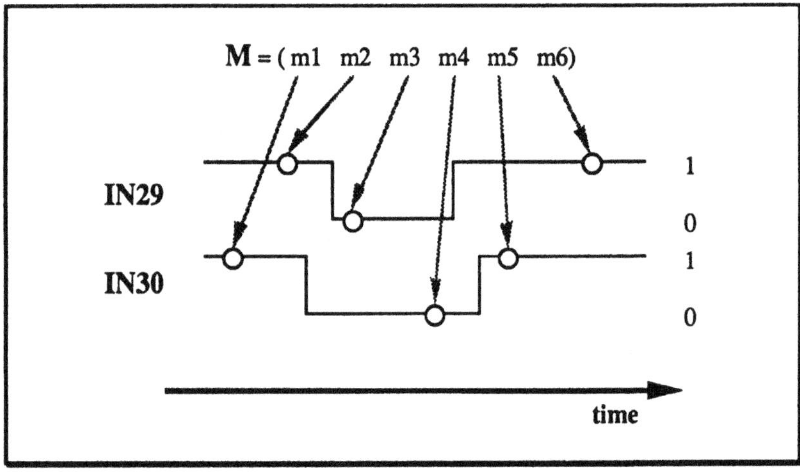

Fig. 35 – M is also compatible with other TDS instances;
this one, for example, specifies a different relative order of the transitions

Hence "M matches S" must not hold. The reason for the dilemma is, of course, that M is too sparse to determine unique temporal relations between the reality-episodes; therefore we cannot decide whether or not they form a model of the TDS time net. Without additional information we are stuck. An analogous problem appears in qualitative measurement interpretation [Forbus83], [Forbus86] where discrete measurements are the only clues about the real behavior and no definitive explanation can be generated as long as nothing is known about the behavior *between* observations.

What then is a form of additional information that eliminates the problem *and* is available in realistic domains? Three possibilities come to mind:

- *Persistence*: We are given the extra information that the value of a quantity is likely to remain the same *after* a measurement. The information is probabilistic: the more recent the measurement, the higher the probability that the value has not changed since then.
- *Granularity*: Quantities change their values only so often in any given bounded interval and there is a non-zero lower bound on the time lapse between two consecutive changes. Within this interval the value remains constant with probability 1; we do not know, however, *where* in the interval we have observed the quantity.
- *No external influence*: If a quantity is adjustable and there is no feedback influence on it, it is guaranteed to keep its value between adjustments.

All three types of knowledge can be expected to be available in a technical domain, e.g. we might know the rate of change for a certain quantity and calculate from this the minimal time it spends within a given interval (= within a qualitative value). We will use a mixture of these concepts: our approach will be based on the existence of a lower bound on the granularity of TDS instances. Complementary we will define the granularity of measurement sequence as a measure of its density. If a measurement sequence is sufficiently dense (its granularity is sufficiently small) w.r.t. the granularity of a TDS instance, we can then be sure not to overlook interesting value changes.

DEFINITION 17: The *granularity of a TDS instance* $\langle Q,H,D\rangle$ is defined as

$$\mathrm{gran}(\langle Q,H,D\rangle) := \min \{ |t - t'| \mid t, t' \text{ appear as end points of episode models in } D, t \neq t' \},$$

i.e. every state within the instance during which *all* quantities remain constant lasts for at least $\mathrm{gran}(\langle Q,H,D\rangle)$.

DEFINITION 18: The *granularity of a measurement sequence* $M=\{\langle q_i,t_i,v_i\rangle\}_{i=1,\ldots,n}$ is the longest gap between any two consecutive measurements of the same quantity in M:

$$\mathrm{gran}(M) := \max \{ t - t' \mid \langle q,t,v\rangle, \langle q,t',v'\rangle \in M, t > t', \neg\exists \langle q,t'',v''\rangle \in M: t' < t'' < t \}.$$

The significance of these two definitions becomes apparent in the concept of *weak determination*. The next definition formalizes the fact that a measurement sequence can determine a TDS description in the strong sense, if the reality-episodes do not become too short, i.e. if occurrences of TDS instances with sufficiently large granularities *exist at all*.

DEFINITION 19: A measurement sequence M *weakly determines* the TDS description $S=\langle Q,H,C\rangle$, iff the following holds:

[M has been observed in Int(M) $\wedge$

$\exists$ TDS description $S'=\langle Q,H',C'\rangle$: $\exists$ TI' instance of S':

TI' occurs in Int(M) $\wedge$ gran(TI') $\geq 2 \cdot$ gran(M)]

$\Rightarrow$

$\exists$ TI instance of S: TI occurs in Int(M).

We now reach the main result of this section: a definition of "matches" that satisfies versions of the completeness and soundness properties where "M determines S" has been replaced by "M weakly determines S".

DEFINITION 20: A measurement sequence $M=\{\langle q_i,t_i,v_i\rangle\}_{i=1,\dots,n}$ *matches* a TDS description $S=\langle Q,H,C\rangle$, iff there is an instance $\langle Q,H,D\rangle$ of S that satisfies the following conditions:

(i) BASIS($\langle Q,H,D\rangle$) = Int(M).

(ii) $\forall \langle q, t, v\rangle \in M: \exists \langle I,v\rangle \in H_q: t \in [I^- ; I^+]$

(all measurements are compatible with the instance)

(iii) $\forall q \in Q: \forall \langle I, v\rangle \in H_q: \exists \langle q, t, v\rangle \in M: t \in [I^- ; I^+]$

(there is at least one observation for each episode)

(iv) $\forall q, q' \in Q; \forall E = \langle I,v\rangle \in H_q, E' = \langle I',v'\rangle \in H_{q'}$:

$[C_{E,E'} \subseteq \text{UNCONSTRAINED} \setminus \{<, m\}$

$\Rightarrow \exists \langle q, t, v\rangle, \langle q', t', v'\rangle \in M: I'^- \leq t' < t \leq I^+]$

(if C requires E to end after E' starts, then there is an observation for E at a point ***after*** *the observation of E')*

(v) $\forall q, q' \in Q: \forall E = \langle I,v\rangle \in H_q, E' = \langle I',v'\rangle \in H_{q'}$:

$[\exists \langle q, t, v\rangle, \langle q', t', v'\rangle \in M: I'^- \leq t' < t \leq I^+ \Rightarrow \neg\, C_{E,E'} \subseteq \{<, m\}]$

(if there is an observation for E at a point ***after*** *the observation of E', C must not require E to end before E' starts.)*

The next two theorems together correspond to the claim that "matches" as defined in definition 20 is both weakly complete and weakly sound. Notice that both proofs depend critically on the fact that the constraints in the TDS descriptions are convex relations.

THEOREM 6 (weak completeness):

$\forall$ measurement sequences M: $\forall$ TDS descriptions S:

M weakly determines S $\Rightarrow$ M matches S.

PROOF: We prove the equivalent claim:

$\forall$ M: $\forall$ S: $\neg$ M matches S $\Rightarrow$ $\neg$ M weakly determines S.

By definition 20 $\neg$ M matches S $\Rightarrow \neg \exists$ TI instance of S: (i) $\wedge$ (ii) $\wedge$ (iii) $\wedge$ (iv) $\wedge$ (v).

$\Rightarrow \forall$ TI instance of S: $\neg$ (i) $\vee \neg$ (ii) $\vee \neg$ (iii) $\vee \neg$ (iv) $\vee \neg$ (v).

Similarly, by definition 19

[M has been observed in Int(M) $\wedge$

$\wedge \exists$ S'=$\langle$Q,H',C'$\rangle$: $\exists$ TI' instance of S':

TI' occurs in Int(M) $\wedge$ gran(TI') $\geq 2 \cdot$ gran(M))

$\wedge \neg \exists$ TI instance of S: TI occurs in Int(M)]

$\Rightarrow \neg$ M weakly determines S.

Abbreviate [...] as X.

We therefore have to prove each of the claims:

($\forall$ TI instance of S: $\neg$ (i)) $\Rightarrow$ X,

...

($\forall$ TI instance of S: $\neg$ (v)) $\Rightarrow$ X.

Case (i):

This case holds vacuously, because for every TDS description S there are infinitely many instances TI with BASIS(TI) = Int(M).

Case (ii):

$\forall$ TI instance of S: $\neg$ (ii) $\Rightarrow$

$\forall$ TI instance of S: $\exists \langle q, t, v\rangle \in$ M: $\forall \langle I,v\rangle \in H_q$: $t \notin [I^- ; I^+]$.

This means that the measurement sequence contains an observation of an episode that does not appear in the TDS description S, hence $\neg \exists$ TI instance of S: TI occurs in Int(M).

Case (iii):

$\forall$ TI instance of S: $\neg$ (iii) $\Rightarrow$

$\forall$ TI instance of S: $\exists q \in Q$: $\exists \langle I, v\rangle \in H_q$: $\forall \langle q, t, v\rangle \in$ M: $t \notin [I^- ; I^+]$.

This means that the TDS description contains an episode for which there is no observation in the measurement sequence. Take an arbitrary instance TI of S with BASIS(TI) = Int(M). Combine the offending episode with one of the adjoining episodes; the resulting episode inherits the value from the adjoining episode. This new TI' is not an instance of S, yet gran(TI') $\geq$ gran(TI). Furthermore M is compatible with

TI'. But if TI' occurs in Int(M), then certainly no instance of S occurs in Int(M).

Case (iv):

$\forall$ TI instance of S: $\neg$ (iv) $\Rightarrow$

$\forall$ TI instance of S: $\exists$ q, q' $\in$ Q; $\exists$ E = $\langle I,v \rangle \in H_q$, E' = $\langle I',v' \rangle \in H_q$':

$$C_{E,E'} \subseteq \text{UNCONSTRAINED} \setminus \{<, m\} \wedge$$
$$\neg \exists \langle q, t, v \rangle, \langle q', t', v' \rangle \in M: I'^- \leq t' < t \leq I^+.$$

This means that there is no evidence in the measurement sequence for E ending after E' starts. It follows that necessarily $I^+ - I'^- < 2 \cdot \text{gran}(M)$, because otherwise two suitable measurements would exist, and therefore $\forall$ TI instance of S: gran(TI) < 2 · gran(M). This contradicts $\exists$ S': $\exists$ TI' instance of S': TI' occurs in Int(M) $\wedge$ 2 · gran(TI') $\geq$ gran(M), hence case (iv) is impossible.

Case (v):

$\forall$ TI instance of S: $\neg$ (v) $\Rightarrow$

$\forall$ TI instance of S: $\exists$ q, q' $\in$ Q: $\exists$ E = $\langle I,v \rangle \in H_q$, E' = $\langle I',v' \rangle \in H_q$':

$$\exists \langle q, t, v \rangle, \langle q', t', v' \rangle \in M: I'^- \leq t' < t \leq I^+ \wedge C_{E,E'} \subseteq \{<, m\}.$$

Here the measurements indicate that E ends only after E' has started, although C requires that it must not. If $I'^- < I^+$ and at the same time $C_{E,E'} \subseteq \{<, m\}$, then TI is not an instance of S, again a contradiction. ■

THEOREM 7 (weak soundness): $\forall$ measurement sequences M: $\forall$ TDS descriptions S:
M matches S $\Rightarrow$ M weakly determines S.

PROOF: Let S = $\langle Q,H,C \rangle$. Assume that M has been observed and that an instance TI' of a TDS description S' with gran(TI') > 2 · gran(M) has occurred in Int(M). We have to show that in this case an instance TI of S must also have occurred in Int(M).

Show:

(a) All episodes in H have occurred in Int(M).

(b) No other episode has occurred in Int(M).

(c) The reality-episodes in Int(M) form a model of C.

Proof of (a):

Let $\langle I, v \rangle$ be an episode in H_q, $H_q \in$ H. Then (iii) implies that there is a measurement $\langle q, t, v \rangle \in$ M so that t $\in$ Int(M). Hence the episode must occur during Int(M).

Proof of (b):

Assume that an instance TI' of a TDS S' = $\langle Q, H',C' \rangle$ occurs in Int(M) which contains an episode $\langle I',v' \rangle \in H'_q \in$ H' and $\neg \exists \langle I,v' \rangle \in H_q \in$ H. Then

(*) $I'^+ - I'^- \geq$ gran(TI') > 2· gran(M),

because otherwise we would have violated the initial assumption. But

$$(*) \Rightarrow \exists \langle q,t,v' \rangle \in M: I'^- \leq t \leq I'^+ \overset{(ii)}{\Rightarrow} \exists \langle I,v' \rangle \in H_q \in H$$

which is a contradiction.

Proof of (c):

(a) and (b) together guarantee that there is a 1-1 mapping between the episodes in H and the reality-episodes during Int(M). We still have to show that reality-episodes obey the temporal constraints in C, i.e. they form a model of N(S). Assume that instead there is a TDS description S' = ⟨Q,H,C'⟩ (same histories as S, but different constraints) and that an instance TI' of S' occurs in Int(M). We show indirectly that TI' is also an instance of S. Since all relations in C are convex we can use the translation defined in section 3.2.5 to convert each $C_{E,E'}$ into a conjunction of end point orderings. If TI' were an instance of S' but not of S, there would have to be at least one literal in the end point form of C that is violated in TI'. Without loss of generality we can assume that this literal relates the common end point of episodes A and B to the common end point of episodes C and D.

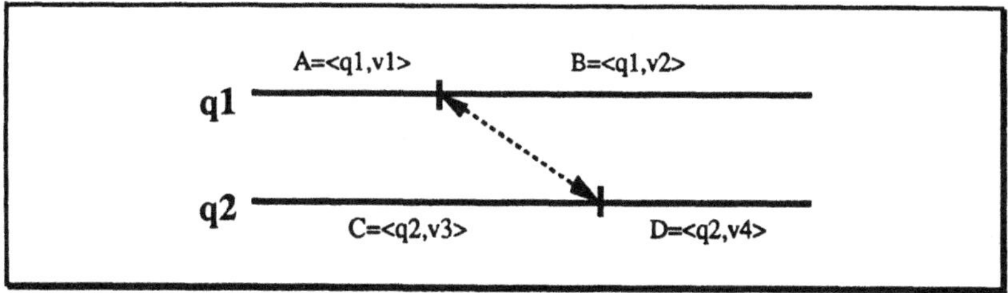

There are 6 cases:

Relation between B^- and C^+ according to $C_{B,C}$	Relation between B^- and C^+ in TI'	
=	<	In TI' $B^- - C^+ \geq gran(TI') > 2 \cdot gran(M) \Rightarrow \exists \langle q, t, v\rangle, \langle q', t', v'\rangle \in M$: $B^- \leq t' < t \leq C^+ \overset{(v)}{\Rightarrow} \neg\, C_{C,B} \subseteq \{<, m\}$. If on the other hand $C_{B,C}$ implies $B^- = C^+$, then $C_{B,C} = \{mi\}$ and therefore $C_{C,B} = \{m\}$ which is a contradiction.
=	>	analogous
≥	<	TI' again implies $\neg\, C_{C,B} \subseteq \{<, m\}$. If on the other hand $C_{B,C}$ implies $B^- \geq C^+$, then $C_{B,C} = \{>, mi\}$ and therefore $C_{C,B} = \{<, m\}$ which is a contradiction.
≤	>	analogous

>	≤	If $C_{B,C}$ implies $B^- > C^+$, then $C_{B,C} = \{>\}$ and therefore $C_{A,D} \subseteq$ UNCONSTRAINED \ {<, m}. It follows from (iv) that $\exists \langle q, t, v\rangle, \langle q', t', v'\rangle \in M$: $C^+ = D^- \leq t' < t \leq A^+ = B^-$ which is a contradiction to $C^+ \geq B^-$ in TI'.
<	≥	analogous

This completes the proof. ■

Before we go on to the next section where we turn the formal definition of "matches" into an algorithm we discuss some of the implications of definition 20.

Depending on the size of a TDS description a fairly large number of elementary conditions have to be satisfied for the match to succeed. Given the possibilities of noisy data and of accidentally missing a crucial observation a match can fail more or less undoubtedly. Should we not differentiate between a patent mismatch and a case where all elementary conditions except for a small fraction are satisfied? In the design of the representation language and definition 20 we considered several possible solutions to this "near-match" problem. There are basically two different approaches:

- *Vague matching*: Instead of defining "matches" as a predicate in the standard sense which is always either true or false, use a fuzzy predicate [Zadeh79] that ascribes to each pair of a measurement sequence and a TDS description a number between 0 and 1 which is a measure of the degree of match.

- *Vague specifications*: Provide constructs in the language for TDS descriptions which allow partial information about values or temporal relations to be expressed. Use an all-or-nothing definition of "matches" on these partially specified TDS descriptions.

On the surface vague matching seems more attractive because in writing down TDS descriptions one does not have to worry about specifying in each case which relations are mandatory and which are less significant. This is all supposedly taken care of once and for all in the matching machinery. As a matter of fact, there is no straight-forward definition of a fuzzy "matches" predicate that produces satisfactory results in all cases. There is no universal criterion for weighing the importance of one temporal relation against the other; the decision must always be based on the background knowledge of what the individual TDS description is intended to capture. Furthermore, it is most unclear how a meaningful interpretation could be given to the numerical result of a match.

For these reasons we prefer the vague specification approach. Apart from allowing episodes to be labelled with more than one value, disjunctive temporal relations are the principal means of vague specification. Consider e.g. the 1-0-transitions in our standard example and imagine that most of the time the two transitions follow each other very closely. The effect could be that the match often fails because the two measurements between the transitions required by condition (iv) of definition 20 cannot be made in the short time.

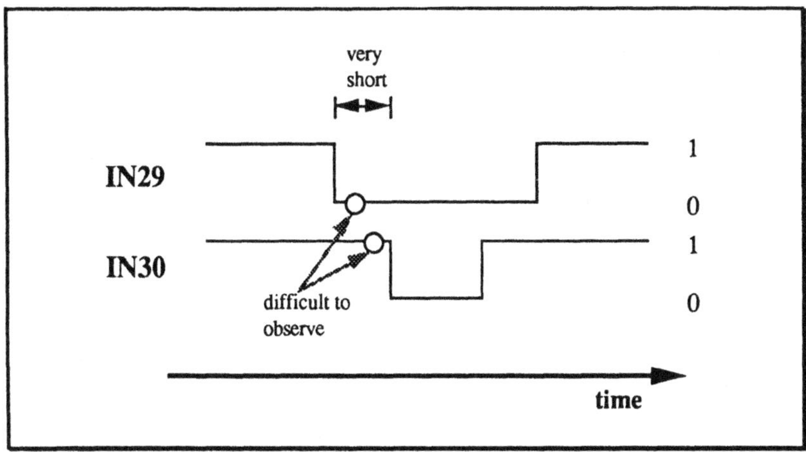

Fig. 36 – A hazard for successfully matching the IN29/IN30 TDS description

Instead of letting a generic matching definition handle the situation, the TDS description language gives more flexibility. One can either

- decide that the overlap is a salient[24] element: in this case the TDS description remains the same and one must try to improve (reduce) the granularity of the measurement sequences; or
- relax the temporal relation specifying the overlap: if e.g. the positions of the 1-0-transitions are insignificant as compared to the 0-1-transitions, one may want to specify the relation between IN30 = 1 and IN29 = 0 as {<, m, o} instead of just

[24] "Salience" cannot be defined within our formal framework. It is determined by the user's intention to concisely capture the relationship between a TDS and a diagnosis.

{o}. The offending elementary condition produced by condition (iv) of definition 20 is thus eliminated and the match is successful; or,

- drop the TDS altogether: if the overlap is *both* significant *and* extremely hard to verify, then this TDS may simply be a bad choice and it may be better to look for alternatives.

We now prove a result about the relationship between "M matches S" and "M is compatible with a particular instance of S". This result will be of importance in section 4.2.3.2 where we have to decide algorithmically whether a measurement sequence is compatible with the occurrence of a TDS instance.

THEOREM 8: Let M be a measurement sequence and TI be an instance of TDS description $S = \langle Q, H, C \rangle$ s.t.

(1) gran(TI) $\geq$ 2 gran(M) and

(2) Int(M) = BASIS(TI).

Then

M is compatible with TI $\Leftrightarrow$ M matches S.

PROOF: $\Leftarrow$: trivial.

$\Rightarrow$: We prove that the instance of S required by definition 20 exists and that it satisfies conditions (i) – (v) of this definition by showing that TI is such an instance.

ad (i): precondition (2) of theorem.

ad (ii): implied by "M is compatible with TI".

ad (iii): implied by precondition (1) of theorem.

ad (iv): implied by precondition (1) of theorem.

ad (v): If TI is an instance of S, then for any two episodes $E = \langle I,v \rangle \in H_q$, $E' = \langle I',v' \rangle \in H_{q'}$ the following holds: $E^+ > E'^- \Rightarrow \neg\, C_{E,E'} \subseteq \{<, m\}$. Together with "M is compatible with TI" this implies (v). ■

We close the section by taking a second look at the measurement sequence M of fig. 34. Comparing it to definition 20 we find that it satisfies (i) – (iii) and (v), but not (iv). By adding two additional measurements for IN30 = 1 and one for IN29 = 0 we get the measurement sequence M' in fig. 37 which indeed matches the situation.

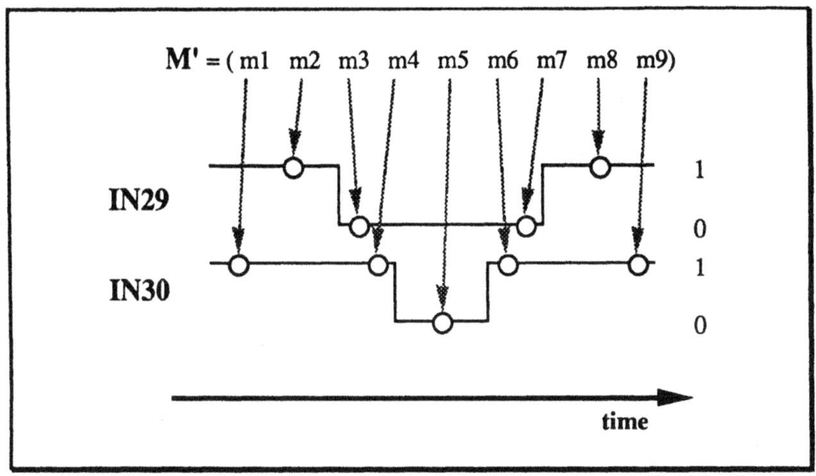

Fig. 37 – A measurement sequence that matches the IN29/IN30 TDS description

3.6 A Temporal Matching Algorithm

3.6.1 Goals

In which sense precisely does the previous section help us in detecting an occurrence of a TDS instance? Definitions 19 and 20 answer three questions:

(1) Which measurements *at least* have to be present in a measurement sequence that weakly determines a given TDS description?

(2) Which measurements are *definitely not* allowed in a measurement sequence that weakly determines a given TDS description?

(3) Down to which granularity does a given measurement sequence determine TDS instances?

All three aspects are essential to the solution of the temporal matching problem. They may, in fact, *be* the solution in the special case where a measurement sequence is somehow given to the temporal matcher. In our scenario, however, no observations are made automatically so that we not only have to match a measurement sequence against a TDS, but we have to develop a constructive interpretation for definition 20 that enables us to *find* a suitable measurement sequence in the first place.

3.6.1.1 Economical Measurement Sequences

As the formulation suggests, the answer to questions (1) and (2) are not complementary to each other, implying that there are different measurement sequences matching the same TDS description. These sequences do not only differ in the time points of their measurements; in addition to the observations required by (1) others may be present which are neither necessary nor contradictory. In some cases even the order of some of the measurements may be insignificant. Although all of these sequences carry the same information w.r.t. the TDS description, by pragmatic standards they are not equivalent. In general observations cause an expenditure of effort and time which depends on a variety of factors such as the accessibility of the quantity to be measured, the skill of the observer, the availability of measuring equipment etc. Rather than going into the details of determining individual costs[25] we assume that they are non-zero so that in principle a rational observer tends to minimize observations unless this optimization is in itself too expensive. While the observation cost optimization thus tends to favor sparse measurement sequences, we may still prefer a denser sequence because of its finer granularity and hence its superior discrimination power.

The algorithm that we present solves this conflict by separating aspects (1) and (2) from aspect (3). For the purpose of suggesting the next observation we ignore the granularity issue and suggest only a minimal sequence consisting of the observations required by (1). If the resulting gaps between measurements of the same quantity become too large, they may be filled arbitrarily with additional measurements until the desired granularity is reached. Notice, however, that despite this separation the matching algorithm provides information about necessary measurements due to aspect (3): from the durations between observations of different episodes it is possible to calculate an upper bound on the granularity of the TDS instance that is currently occurring.

3.6.1.2 Interleaving Planning and Matching

Another important design aspect concerns the combination of planning and matching. Even if we had algorithms for planning a measurement sequence and for matching it against a TDS description, it would be wasteful to run the two in sequence. An initial segment of the measurement sequence is often sufficient for a mismatch; in such a case

---

[25] For a discussion of test selection based on test cost see [Schuch89].

the rest of the measurements would have been carried out needlessly and, strictly speaking, even planning it would have been in vain. If the implementation of the planning algorithm can be made fast enough to permit on-line operation, it is better, of course, to interleave planning and matching so that further measurements are planned only after it has been checked that the initial segment of the sequence obtained so far can still be extended to a sequence that matches the TDS description[26]. The algorithm therefore alternately suggests the next measurement and matches the actual observation against the prediction. Since only the obligatory measurements are ever suggested and additional "voluntary" measurements may be needed to lower the granularity of the sequence, we do not require that the quantities in the suggestion and the actual observation are identical. Apart from gaining flexibility by being able to adjust the suggestion depending on the new observation, as an added benefit this solution provides an interface to monitoring: sensor data which arrive at regular intervals can be treated just like "voluntary" measurements and thus contribute to the matching process at the earliest possible time. That way our framework is general enough to be useful in every conceivable mixture from purely demand-driven observations to pure monitoring.

3.6.2 Overall Design of the Algorithm

Taking into account all these goals we can develop a first sketch of the algorithm (fig. 38). The initial input is the TDS descriptions whose occurrence we want to detect. No observations have yet been made meaning that we have information about what is happening. We now perform a three-step loop. Based on the TDS description and the observations made so far we compute a set of quantities one of which should be measured next. As we will see in one of the examples given in section 3.6.6, there may be more than one suggestion depending on the form of the TDS description. In the next step, which may or may not be under program control, we select a quantity (not necessarily one from the suggested set) and perform the measuring action. Finally, we match the actual observation against the TDS description and the context formed by the measurements obtained so far.

With each loop the accumulating measurements reduce the set of TDS instances that the growing measurement sequence is compatible with.

---

[26] This is, of course, the standard technique of manually applying the lazy evaluation strategy to a generator-consumer pair.

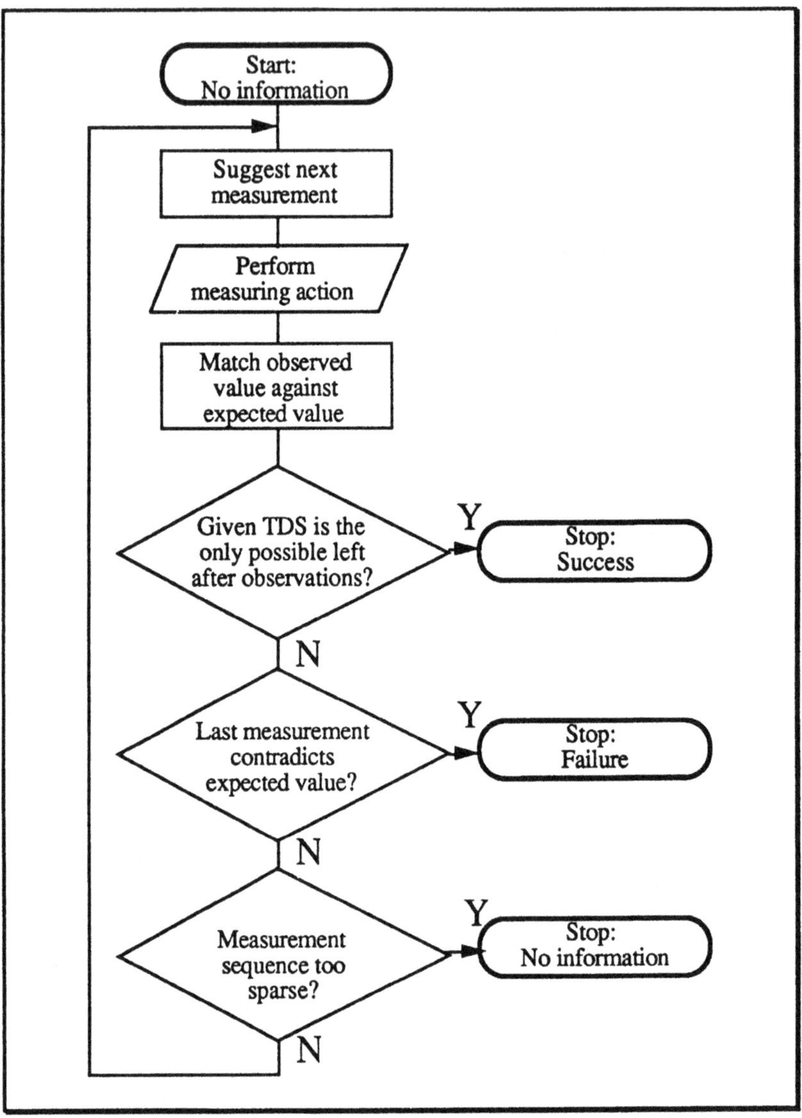

Fig. 38 – High-level flow diagram of the matching algorithm

The cycle can be left in three different ways:

- The information accumulated so far may be sufficient to weakly determine the given TDS description; in this case we report success. Whether this implies that an instance of the TDS has actually occurred, depends on the granularities of the measurement sequence and of the possible TDS instances.

- The latest measurement may contradict the expected value(s) computed from the TDS description and the previous observations; in this case we report failure.

- As shown in fig. 38, we may be left in a third case which is typical of the problems in temporal matching. If, by accident, the measurement sequence has not been dense enough, we may have missed the chance to verify a temporal constraint (e.g. the overlap between the first episode of IN20 and the second episode of IN30 in our example TDS). Once this has happened there is nothing we can do short of starting all over again. We have not observed a contradiction nor can we be sure that the behavior is an instance of the given TDS description, so we abort the match with result `unknown`.

Notice that the last case does not depend on the particular granularity that we want to achieve; instead one of the measurements required by conditions (iii) and (iv) of definition 20 was not produced in time, hence the resulting measurement sequence cannot weakly determine the given TDS description *at all*!

3.6.3 The Reality Time Net and the Observation Rule

Looking again at fig. 38 we can see that the most important question is how to organize the monotonically growing information drawn from the observations. In the course of the matching process we make use of this information in the following ways:

- Together with the TDS description it forms the basis for suggesting the next measurement.

- Similarly it is used to compute the predictions against which the actual observations are matched.

- As a side effect of the matching step, the latest measurement is incorporated into it for the next time through the loop.

Recall that we want to verify (i) that exactly the episodes mentioned in the TDS description actually occur and (ii) that their relative positions satisfy the temporal constraints. We will use two separate data structures to keep track of the information pertinent to (i) and (ii).

Treating (i) is relatively easy: the progress of the matching process can be represented using a sweep-line technique. At any point of the process imagine the set of episodes to be partitioned into three classes:

- `sleeping`: episodes that have not yet been observed;
- `open`: episodes which have been observed at least once, but whose successor episodes have not yet been observed;
- `closed`: episodes which (directly or via transitivity) precede an episode in `open`.

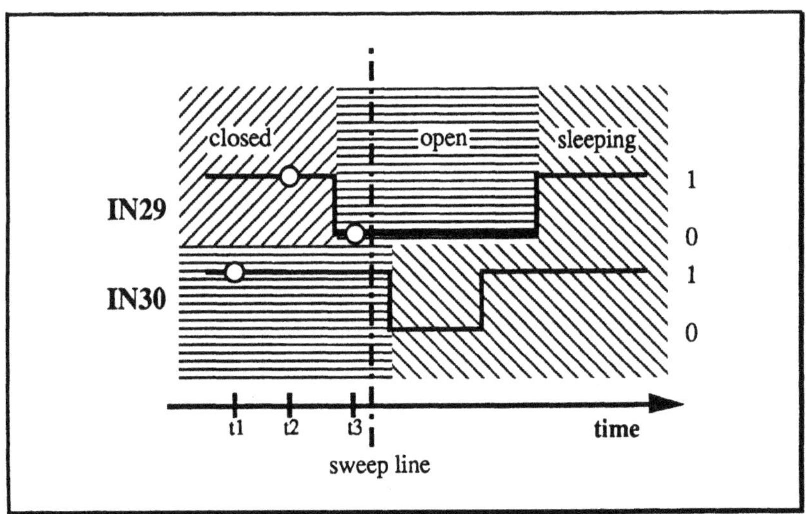

Fig. 39 – The sweep-line after the first three measurements

Fig. 39 gives an example. At the beginning the sweep-line is "left" of all episodes, i.e. all episodes are in `sleeping`. During the line is swept across the TDS description from "left" to "right": the diagram shows e.g. which episodes belong to `sleeping`, `open` and `closed` after the measurements ⟨IN30, t_1, 1⟩, ⟨IN29, t_2, 1⟩, ⟨IN29, t_3, 0⟩. The position of the sweep-line is updated each time a measurement is associated with an episode that is not already in `open`. If all goes well, the sweep-line will end up at the "right margin" of the TDS description indicating that `sleeping` is empty and that therefore all episodes have been corroborated by observations. The use of the sweep-line for planning and matching is discussed in sections 3.6.4.2 and 3.6.4.3, respectively.

As far as (ii) is concerned we have to represent information about the temporal relations among a collection of A-intervals (the reality-episode intervals). Each additional measurement contributes constraints on some of these relations and via transitivity modifies others until we can decide whether or not the relations between the reality episodes satisfy the temporal constraints between the corresponding episodes in the TDS description. This is precisely the type of task at which Allen's interval calculus – in particular the time net paradigm and the path-consistency algorithm – is best.

We therefore choose a time net as our representation; since – contrary to the temporal constraints in the TDS description – it contains information about events in the "real world", we call it a *reality time net.*

The initial state in which we hypothesize that the reality-episodes mentioned in the TDS description are present but know nothing about their temporal relations[27] can be represented as an unconstrained reality time net.[28]

Similarly, a match is successful, if after all observations have been made and the reality time net has been modified accordingly the relations that remain possible between the reality-episodes are subsets of the respective TDS constraints. It is important to note that it is not necessary (and therefore not economical) to nail down the relation between two reality-episodes to a single primitive relation. If a temporal constraint in the TDS description specifies a disjunctive relation, then we may stop as soon as the actual relation has been narrowed down to a subset of the constraint. We then still do not know the *exact* temporal relation between the two episodes, but whatever it is it fits the TDS description.

The question remains how to capture the idea of "observations constraining temporal relations between episodes". The answer is connected to the role that the instance of the TDS description plays in definition 20. Recall that the definition requires the existence of a TDS *instance* with certain properties. It is, however, easy to see that we can guarantee the existence of the instance without constructing it explicitly. If the measurements taken can be mapped to the episodes of the TDS description in such a way that the conditions of the definition are satisfied (e.g. each observation is mapped to exactly one episode, each episode has at least one observation mapped to it, etc.),

---

27 The presence as such is verified using the sweep line technique.

28 In section 3.6.4.1 we will explain why we have to start with a slightly different time net.

then it is trivial to pick a model of the temporal constraints that is compatible with the mapping: simply assign any point between the last observation mapped to episode E and the first observation of its successor episode E' to both E^+ and E'^-. Matching an observation against the TDS description therefore means primarily picking an episode and associating the observation with it. Now, a single such assignment does not carry any useful information about the temporal relations involved. But consider a measurement sequence which contains (among others) the observations $m_1 = \langle q_1, t_1, v_1 \rangle$ and $m_2 = \langle q_2, t_2, v_2 \rangle$ where $t_1 < t_2$ (fig. 40). Let E_1 and E_2 be the episodes that m_1 and m_2 are associated with. Then it follows directly that E_2 cannot possibly end before E_1 begins, because m_2 is the later of the two measurements. Translated back into Allen relations this is equivalent to saying that E_2 UNCONSTRAINED \ {<, m} E_1. Using this implication which we call the *observation rule* we can throw out certain interval relations each time a new measurement is added; Allen's propagation algorithm allows us to determine which other relations in the reality time net are also affected. This strategy can be viewed as a kind of Waltz filtering where the observation rule is the only constraint and the labeling is refined incrementally.

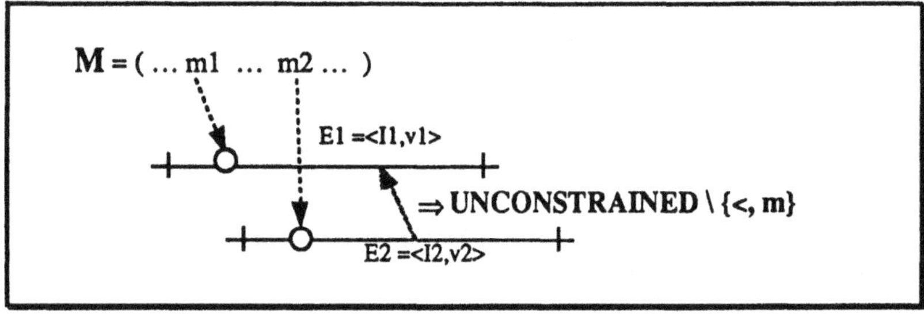

Fig. 40 – The observation rule

Interestingly, both the sweep line and the reality time net/observation rule can be put to use very elegantly in both the planning and the matching step of the algorithm. This will be seen in greater detail in the next three sections where we develop the algorithm.

3.6.4 The Algorithm in Detail

3.6.4.1 MAIN - The Main Loop of the Algorithm

Basically, the main loop of the matching algorithm is a straight-forward translation of the flow diagram in .fig. 38 where the initialization and the final check have been refined according to the chosen data structures. The algorithm is a one-pattern matcher:

as it stands the input consists of *one* TDS description whose occurrences are to be verified. Indications of how the algorithm could be used to test for several TDS descriptions simultaneously are given in the section on its applications in a diagnostic system. The only notable deviation from the outline given in the previous sections concerns the initialization of the reality time net. In order to provide an uncluttered birds-eye view of the reality time net and the observation rule we deliberately glossed over a problem and pretended that only by applying the observation rule and propagating its results we could arrive at any desired labeling. This is not so, as the TDS description in figure 41 shows.

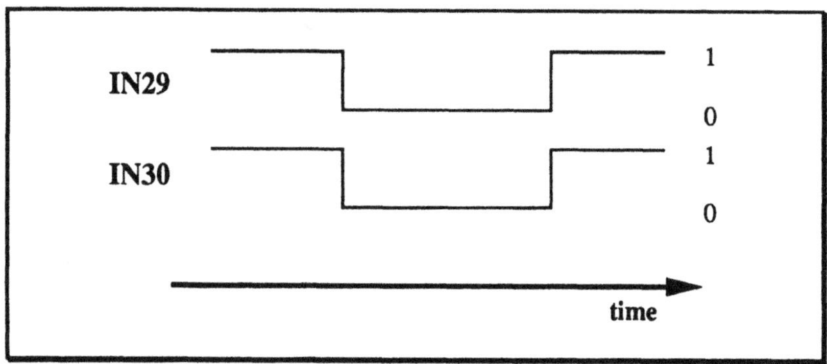

Fig. 41 – A TDS description involving simultaneous episode transitions

The problem here is the simultaneous change of the two quantities. Although condition (iv) of definition 20 ensures that every overlap specified by the temporal constraints is actually present, there is no positive condition which guarantees that simultaneous changes really take place simultaneously. In fact, given the limitations of our measurements there cannot be such a condition. A look at the proof of theorem 7 reveals our solution to the problem: condition (v) of definition 20 is used in the form of a *via negationis* argument; if there were an overlap one way or the other instead of the simultaneous change and our measurement sequence were dense enough, we would be certain to find measurements that corroborate the overlap. Hence, in the absence of conflicting measurements we *assume* that the change was indeed simultaneous[29].

---

[29] This may sound as if temporal matching were going to become non-monotonic through the backdoor. In fact, this is not the case, because we either notice the conflict at once or we will *never again* get information that could invalidate our assumptions.

The dilemma reappears when we apply the observation rule to the example.

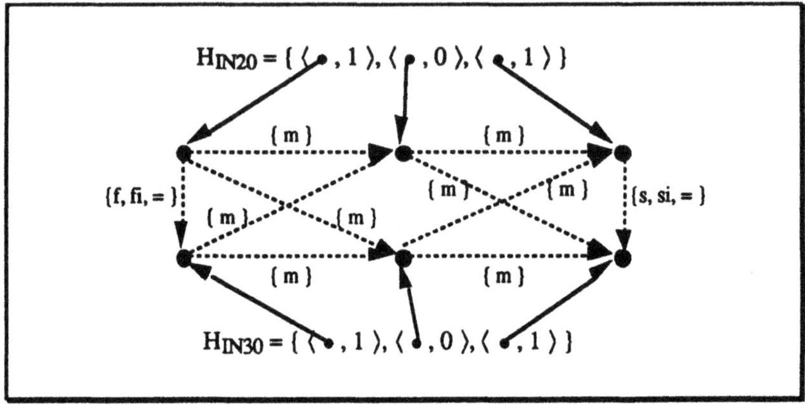

Fig. 42 – Some temporal constraints corresponding to fig. 41

As figure 42 shows, most of the temporal constraints are {m}. Unfortunately, an application of the observation rule always deletes *both* < and m, never only one. There is thus no way to go from UNCONSTRAINED to {m} using the observation rule only[30]. An analogous argument applies to {<, m} constraints. There is only one conceivable solution: initialize those edge labels in the reality net with UNCONSTRAINED which can be narrowed down by the observation rule; initialize the others (which are {m} and {<, m} with their temporal constraints from the TDS description and see if the measurement sequence contains contradictory observations.

```
Algorithm MAIN:

Purpose: Detect an occurrence of a given TDS by planning an
appropriate measurement sequence and incrementally matching the
actual observations against the TDS description.

Input: A TDS description S = ⟨Q,H,C⟩.

Output: The result of the match, one of
      success: a measurement sequence was observed that matches S;
      failure: the measurement sequence contradicts S;
      insufficient data: the measurement sequence was too sparse.
      In case of success M contains a measurement sequence that
      matches S.
```

30 This is true no matter which relation we try to constrain first.

```
Temporary data structures:
      sleeping, open, closed: sets of episodes that mark
            the position of the sweep line;
      G = ⟨V,R⟩: a time net (the reality time net);
      M: an association list of pairs ⟨m,E⟩ where m is a measurement
            and E is an episode from H;
      suggestions: a set of quantities.

begin

(1)   Initialize reality time net G:
      • set of intervals V ← the set of episodes in H
      • labeling function R: for each pair E, E' episodes in H:
            R_E,E' ← UNCONSTRAINED;
(2)   for each pair E, E' episodes in H:
        if C_E,E' = {m} or C_E,E' = {<, m}
            then R_E,E' ← C_E,E' and propagate in the reality
                                   time net according to Allen;
(3)   open ← Ø;
(4)   sleeping ← the set of episodes in H;
(5)   closed ← Ø;
(6)   M ← the empty sequence
(7)   suggestions ← SUGGEST(S,G,open,sleeping,closed);
(8)   while suggestions ≠ Ø do
(9)         obtain next measurement(m);
(10)        result ← MATCH(m,S,G,open,sleeping,closed);
(11)        if result ≠ success,then report result, stop;
(12)        suggestions ← SUGGEST(S,G,open,sleeping,closed);
(13)  end;
(14)  if for all pairs E, E' episodes in H: R_E,E' ⊆ C_E,E'
            then report success
            else report insufficient data;
end.
```

Comments:

(1) see text.

(2) Because of step (14) we keep G in minimal labeling form at all times. At the beginning we must therefore propagate the consequences of the non-UNCONSTRAINED labels.

(3)-(5) Initialize sweep line to be "left" of all episodes.

(6) No measurements have been taken yet.

(7) Compute a set of quantities one of which should be measured next.

(8) In particular sleeping ≠ Ø implies suggestions ≠ Ø.

(9) Not necessarily one of the measurements suggested. Not necessarily under program control.

(14) This test relies critically on the completeness of Allen's algorithm for the minimal labeling problem for convex relations. Theorems 4 and 5 guarantee that the labels $R_{E,E'}$ are indeed minimal so that the test cannot fail because the labels in the reality time net are not subsets of the temporal constraints, although in fact the temporal relations between the reality episodes satisfy the constraints.

3.6.4.2 SUGGEST - Planning the Next Observation

As we have explained before, it is not sensible to prescribe a unique measurement at every point in the matching process. Depending on the temporal constraints there may be several possible next measurements and even if this is not the case, the user of the system may select a different quantity for the next measurement because he wants to minimize the granularity of the sequence or simply because certain sensor readings happen to be available at that moment. Nevertheless it is not always easy to determine which measurements are useful in a given situation and the user may want to be guided through the measurement sequence by the matching algorithm.

The second motive behind SUGGEST is the test in line (8) of the MAIN algorithm: an empty set of suggestions signals that no further measurements can increase the information about the reality-episodes and that we are ready to carry out the final test of the reality time net against the temporal constraints.

How do we go about suggesting measurements? Reasons to observe a quantity can come from two sources:

- Not all of the episodes in the quantity's history have been observed yet.
- The relations between the episodes in the quantity's history and other episodes have not yet been reduced to their corresponding constraints.

Both of these sources can be reinterpreted in terms of the temporary data structures. Episodes in `sleeping` are automatically candidates for being observed, because they have not yet been associated with a measurement. Figuratively speaking, they are between the sweep-line and the right margin of the TDS description and the only way of driving the sweep-line to the right margin is to observe all episodes on the way (i.e. shifting them into `open` and eventually into `closed`). On the other hand temporal relations in the reality time net can only be modified through applications of the observation rule. Since its primary effect is very specific (remove < and m from one relation in the net), we can look for relations in the current reality time net which differ from their target in exactly this way. If we find such a pair of episodes E, E' s.t. $\{<,m\} \cap R_{E,E'} \neq \emptyset$ and $\{<,m\} \cap C_{E,E'} = \emptyset$, then we can explicitly plan to eliminate the unwanted relations by observing first E', then E, and applying the observation rule.

Both sources together usually produce a rather large set of suggested episodes. Fortunately there are a number of rules by which we can eliminate candidates in a second phase of SUGGEST without risking to miss a significant observation. Three of these deletion rules have proved to be very effective in the examples on which we ran the algorithm; each of them will be discussed in turn with the help of a graphical example.

The simplest rule says: "Do not plan ahead too far." If we have two candidate episodes one of which precedes the other according to the TDS description, then we can drop the latter from the candidate set, because the former has to be corroborated first anyway. If in fig. 43 E, E' and E" are all candidates after the first phase of SUGGEST, then E' and E" can be safely eliminated for that reason. There is no danger of skipping them unobserved, since they will be added to the candidate list again the next time around and then E will not be a candidate any more.

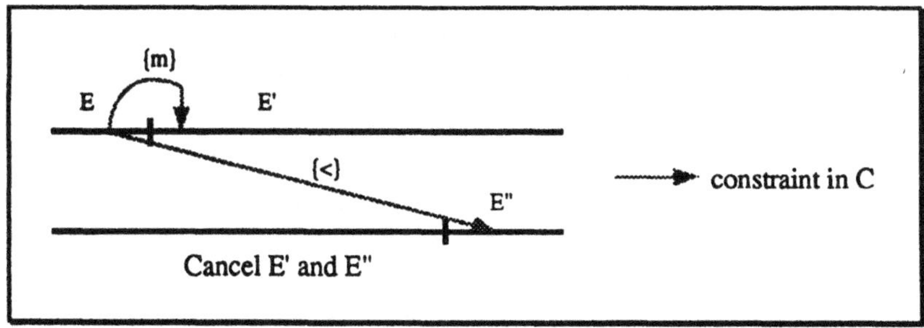

Fig. 43 – The first deletion rule

The second rule is illustrated in fig. 44. Here E is a candidate, although it has been observed before (obviously because it is to be used in an application rule). Since E is being observed a second time, this can only mean that E is the second observation of the couple required by the observation rule. But from $C_{E',E}$ we know that we will have to observe E at least *one more* time (after E") to verify the overlap. But E' is also a candidate which must be observed before E". Hence we can postpone the observation of E until after E" and hence after E'. We may therefore drop E from the candidate list.

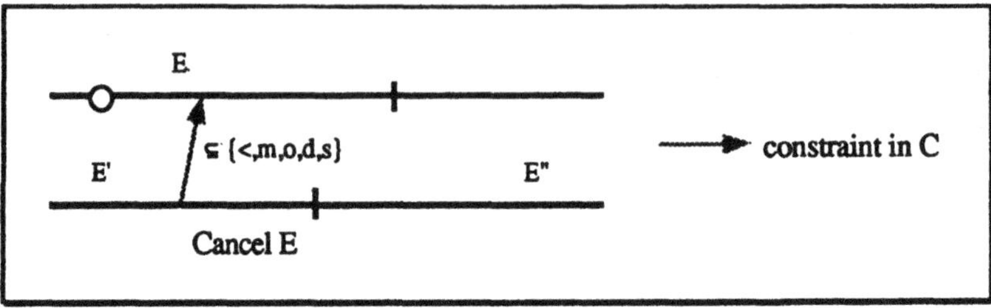

Fig. 44 – The second deletion rule

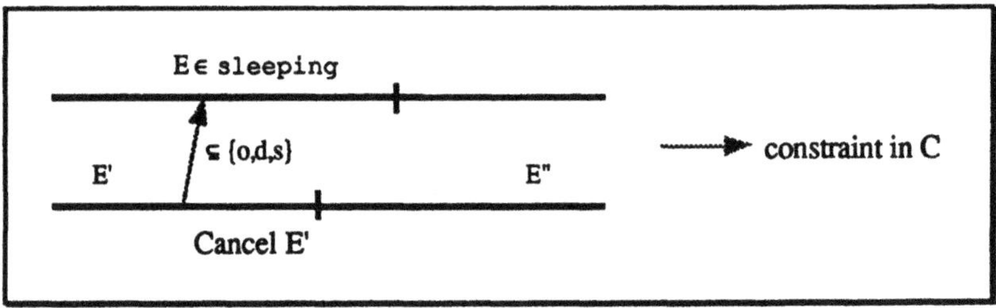

Fig. 45 – The third deletion rule

Fig. 45 shows the pattern for the third rule. From the constraint between E' and E we can conclude that E will have to be observed at least twice to verify the different overlaps. In this situation it is always a good rule to make the first observation of E as early as possible and the second as late as possible, because then they provide maximum leverage for applications of the observation rule. In particular we will have to observe E' (again) after E so we may as well eliminate E' now.

```
Algorithm SUGGEST:

Purpose: Suggest a set of quantities one of which should be measured
next. The suggestion is made on the basis of the conditions in
definition 20. It is not binding, because one might want to volunteer
additional measurements to keep the granularity of the measurement
sequence low.

Input: A TDS description S = ⟨Q,H,C⟩;
      a reality time net G;
      a partitioning of the episodes into open, sleeping and closed;
      an association list M.

Output: A set of quantities.
```

```
Temporary data structures:
      candidates: a set of episodes which should be corroborated
            by observation as soon as possible;
      follow-ups: a set of episodes which - in order to be able to
            apply the observation rule - should be observed after
            one of the episodes in candidates.

begin

Find candidates:
(1)   follow-ups ← Ø; candidates ← sleeping;
(2)   For each pair of episodes E, E' such that
      {<,m} ∩ R_{E,E'} ≠ Ø and {<,m} ∩ C_{E,E'} = Ø do:
(3)         if E' ∈ sleeping
                  then  candidates ← candidates ∪ {E'};
                        follow-ups ← follow-ups ∪ {E};
                  else  candidates ← candidates ∪ {E};

Prune candidates:
(4)   for each E ∈ candidates, E' ∈ (follow-ups ∪ candidates),
      C_{E',E} ⊆ {<,m}:
            delete E from candidates;
(5)   for each E, E' ∈ candidates, E ∈ open, C_{E',E} ⊆ {<,m,o,d,s}:
            delete E from candidates;
(6)   for each E, E' ∈ candidates, E ∈ sleeping, C_{E',E} ⊆ {o,d,s}:
            delete E' from candidates;


(7)   Suggest the quantities to which the episodes in candidates
      belong;
end.
```

Comments:

(1) All episodes that have never been observed are automatically candidates. The list of follow-ups is filled in line (3).

(2) In these cases an observation of E' followed by one of E will lead to the elimination of the unwanted relations via the observation rule.

(3) Here we run into a little technical difficulty. We would like to suggest *two* observations in sequence. According to the logic of the MAIN program loop the second suggestion has to wait until next time around the loop. We keep the second elements in the follow-ups list so that we can use them for pruning in step (4). The problem is alleviated, if the first observation of the pair has already been made (E' ∉ sleeping).

(4)-(6) see text.

(7) The candidates list contains episodes, not quantities as is required by the input-output relation of SUGGEST. Notice that in case of adjustable quantities the suggestion is verbalized as a request for setting the quantity to the episode value. "Measurements" of adjustable quantities generally take precedence over measurable quantities, because user actions drive the system through the TDS.

3.6.4.3 MATCH - Matching Observations Against TDS Descriptions

After everything else has been taken care of the matching itself is surprisingly the simplest part of the algorithm; practically all of the techniques needed have already been described. Basically, four things have to be done:

- We have to look for an episode with which we can associate the measurement. The episode must be found in that part of the corresponding history which is not yet completely in the past.
- Using the sweep-line we have to check whether the gap between the episode associated with the last measurement and the episode chosen is too large. If the measurement sequence is too sparse, there may have been episodes in between which were not corroborated according to condition (iii) of definition 20.
- Next we apply the observation rule and see if any of the results contradict the temporal constraints.
- If no contradictions are found, we update the temporary data structures; the reality time net is modified by the Allen propagation during application of the observation rule; we add the association between the new measurement and its episode to the association list and shift the sweep-line position accordingly.

```
Algorithm MATCH:

Purpose: Match the latest observation against the context formed by
the TDS description and the measurements obtained before.

Input: A TDS description S = ⟨Q,H,C⟩;
      a reality time net G;
      a partitioning of the episodes into open, sleeping and closed;
      an association list M;
      a measurement m = ⟨q,t,v⟩.

Output: The result of the match, one of
      success: a measurement sequence was observed that matches S;
      failure: the measurement sequence contradicts S;
```

insufficient data: the measurement sequence was too sparse.

As a side effect, the temporary data structures are updated.

```
begin
(1)   Find the earliest episode E = ⟨I,v⟩ in H_q \ closed;
(2)   if there is none, then report failure, stop;
(3)   if E ∈ sleeping and there exists E' ∈ sleeping s.t.
      C_E',E ⊆ {<,m},
            then report insufficient data, stop;
(4)   for each pair ⟨m',E'⟩ ∈ M do
(5)         add E  UNCONSTRAINED \ {<,m}  E' to G and propagate
            according to Allen;
(6)         if the resulting G is not globally consistent
                  then report failure, stop;
(7)   if E ∈ sleeping then
            closed ← closed ∪ (open ∩ H_q);
            open ← (open \ H_q) ∪ {E};
            sleeping ← sleeping \ {E};
(8)   Append ⟨m,E⟩ to M;
(9)   Report success;
end.
```

Comments:

(1) The new measurement $\langle q,t,v\rangle$ has to be associated with an episode that has the same value and does not lie in the past (i.e. is a member of closed). If there is more than one episode, we choose the earliest; if we didn't, the match would fail in step (3).

(2) If no episode with the same value can be found, the measurement contradicts the TDS description (condition (ii) of definition 20 is violated).

(3) We are about to move the sweep line further to the right in q's history. Whenever we do this we have to check whether an episode has gone unobserved (i.e. still belongs to sleeping) that completely precedes the new episode. If there is one, we have missed our last chance to observe it and condition (iii) of definition 20 is violated.

(4) These are the observations taken before and their associated episodes.

(5) Apply the observation rule to E and each of the episodes associated with an earlier observation.

(6) Recall that Allen's path-consistency algorithm detects all inconsistencies when only convex interval relations are present (theorems 3 and 5). Inconsistencies appear when the application of the observation rule reduces the label on an edge of G to the empty disjunction; this typically happens when an overlap is detected where

```
none is allowed by the temporal constraints. Notice also that after
propagation G is in minimal labeling form again which is important
for step (14) of MAIN.
```

```
(7) If the episode chosen in (1) has been observed for the first
time, we have to shift the sweep line one episode to the right in the
corresponding history h_q.
```

```
(8) Record that m has been associated with E.
```

3.6.5 Correctness of the algorithm

At the end of this section we prove two results about the temporal matching algorithm which demonstrate that it indeed implements definition 20. Before we do so let us state what exactly we want to prove – or rather, what we *cannot* expect from a correctness proof. Naturally we cannot prove that by following the steps in the algorithm we will always detect an occurrence of the given TDS. But we cannot even prove that if the match turns out successful, we have really observed an occurrence. All we can prove is that in the case of a successful termination the measurements obtained match the TDS description in the sense of our definition. Whether this suffices to actually determine the TDS depends on the granularity of the measurement sequence and our external knowledge about the minimal granularity of TDS instances; the algorithm does not make any statements about these. This is theorem 9. Conversely we are interested in the cases where the algorithm terminates with "failure" or "insufficient data". In particular we want to be sure that the failure is not a temporary condition which can be turned into a successful match just by obtaining some more observations, but that instead we are justified to abort the matching process. This is theorem 10.

THEOREM 9: If the algorithm MAIN terminates with "success" then the measurement sequence in M matches the TDS description S.

PROOF: Assume that the premise holds. Let $\{m_i = \langle q_i, t_i, v_i\rangle\}_{i=1,\ldots,n}$ be the measurement sequence collected in M. Construct an instance $\langle Q,H,D\rangle$ of S as follows: for each episode E in $H_q \in H$ define

$$FIRST(E) := \min \{ t \mid \langle\langle q, t, v\rangle, E\rangle \in M \} \quad \text{and}$$
$$LAST(E) := \max \{ t \mid \langle\langle q, t, v\rangle, E\rangle \in M \}.$$

For each episode E define D as follows:

• if E is initial episode of H_q : let $I^- := t_1$; for I^+ as in otherwise case;

• if E is final episode of H_q : let $I^+ := t_n$; for I^- as in otherwise case;

• otherwise let $Q_1(E) := \{ E' \mid C_{E',E} = \{m\} \}$,

$$Q_2(E) := \{ E' \mid C_{PRED(E),E'} = \{m\} \},$$

$$\text{EARLIEST}(E) := \max_{E' \in Q_1(E)} \text{LAST}(E') \text{ and}$$

$$\text{LATEST}(E) := \min_{E' \in Q_2(E)} \text{FIRST}(E'),$$

pick any $c \in \mathbb{R}$ s.t. EARLIEST (E) $\le c \le$ LATEST (E) and

set $I^- := c$ for all $E' \in Q_2(E)$ and $I^+ := c$ for all $E' \in Q_1(E)$,

where PRED(E) is the episode directly preceding E in the same history. D is a model of N(S), since already the partial information from the measurements suffices to constrain the relations between the reality-episodes to subsets of the temporal constraints.

We now prove that $\langle Q,H,D\rangle$ satisfies conditions (i) – (v) of definition 20.

ad (i): True via construction.

ad (ii): All measurements have been successfully associated with episodes, which implies (ii).

ad (iii): At the end of MAIN `suggestions` = `sleeping` = ∅, i.e. all episodes have had measurements associated with them.

ad (iv): $C_{E,E'} \subseteq \text{UNCONSTRAINED} \setminus \{<, m\} \Rightarrow C_{E,E'} = \{mi\} \vee C_{E,E'} = \{mi, >\} \vee C_{E,E'} = \{>\} \vee C_{E,E'} \cap (\text{UNCONSTRAINED} \setminus \{<, m, mi, >\}) \neq \emptyset$. If $C_{E,E'} = \{mi\} \vee C_{E,E'} = \{mi, >\}$, then initially $R_{E,E'} = C_{E,E'}$. Furthermore there exist $m = \langle q,t,v\rangle$, $m' = \langle q', t', v'\rangle$ s.t. $\langle m, E\rangle, \langle m', E'\rangle \in M$. If $t' > t$, then by the observation rule we would obtain $R_{E,E'} = \emptyset$ which is a contradiction to successful termination. Hence $t' < t$. If $C_{E,E'} = \{>\}$ and m, m' as above, then $t' > t$ implies that $E' \in$ `sleeping` at t. This would have caused termination with "insufficient data", again a contradiction which implies that instead $t' < t$. If $C_{E,E'} \cap (\text{UNCONSTRAINED} \setminus \{<, m, mi, >\}) \neq \emptyset$, then initially $R_{E,E'} = \text{UNCONSTRAINED}$. On the other hand the final $R_{E,E'} \subseteq C_{E,E'}$. This can only happen through application of the observation rule to E' or one of its successors and E or one of its predecessors. Let the respective measurements be $m'' = \langle q', t'', v''\rangle$ and $m''' = \langle q, t''', v'''\rangle$ and let m and m' be as above. Then

$$\underset{\substack{\uparrow \\ \text{E' or} \\ \text{successor}}}{t' \quad \le} \quad \underset{\substack{\uparrow \\ \text{observation} \\ \text{rule}}}{t'' \quad <} \quad \underset{\substack{\uparrow \\ \text{E or} \\ \text{predecessor}}}{t''' \quad \le} \quad t.$$

ad(v): If $m = \langle q,t,v \rangle$, $m' = \langle q', t', v' \rangle$, $\langle m, E \rangle$, $\langle m', E' \rangle \in M$, $t' < t$, then after application of the observation rule $R_{E,E'} \cap \{<, m\} = \emptyset$.
Then $R_{E,E'} \subseteq C_{E,E'} \Rightarrow \neg\, C_{E,E'} \subseteq \{<, m\}$. ■

THEOREM 10: If the algorithm MAIN terminates with "failure" or "insufficient data", then no measurement sequence that contains M as an initial segment matches the TDS description S.

PROOF: Let $\{m_i\}_{i=1,\ldots,n}$ be the measurement sequence collected so far. The idea of the proof is to show that in each of the cases where the algorithm reports "failure" or "insufficient data" the contradiction or missing observation occurs at or before t_n so that no later observation can make up for it.
Case 1 (line (14) of MAIN): Here $R_{E,E'} \not\subseteq C_{E,E'}$ for at least one pair of episodes E, E'. However `suggestions` = ∅ implies that no episodes remain unobserved and no opportunities are left for applying the observation rule. This typically happens when the measurement sequence has been too sparse, but all episodes have nevertheless been observed. An example is shown in figure 35.
Case 2 (line (2) of MATCH): The last measurement m_n directly contradicts the TDS description, because a value has been measured that does not appear at all in the remaining part of the history. This is a hard failure and naturally irreparable.
Case 3 (line (3) of MATCH): If E is the earliest episode with which we can associate m_n, we can never again associate a measurement with E'; if we did, $R_{E,E'} \cap \{<,m\} = \emptyset$ (observation rule) and therefore $R_{E,E'} \not\subseteq C_{E,E'}$. Hence E' must remain forever unobserved, violating condition (iii) of definition 20.
Case 4 (line (6) of MATCH): A globally inconsistent G indicates that there is no instance of the initial reality time net G which is compatible with the observations so far. Since the initial G is a relaxed version of the temporal constraints C, there is also no instance of C which is compatible with the observations. Further observations will not remove the contradiction. ■

3.6.6 Extended Examples: The Algorithm at Work

Although we hope that the comments in the previous section have helped in conveying the basic ideas underlying the temporal matching algorithm, we feel it is best to illustrate its operation with an extended example. In this section we will first walk through a few steps of our familiar example TDS. However, this TDS – compact as it is – does not exercise the matching algorithm to its extremes; therefore we will give a second (fictitious) example which has been constructed for the purpose of demonstrating the influence of extensive use of disjunctive relations on the matching process.

3.6.6.1 First Example

We refer to the episodes in the IN29/IN30 TDS using the numbers in fig. 32. The same figure shows the complete TDS time net associated with the TDS description.

The first step in MAIN is the initialization of the reality time net G which after running Allen's propagation algorithm has the relations in fig. 46. Basically G contains no information apart from which episodes follow which within the same history.

to / from	1	2	3	4	5	6
1	=	m	<	?	?	?
2	mi	=	m	?	?	?
3	>	mi	=	?	?	?
4	?	?	?	=	m	<
5	?	?	?	mi	=	m
6	?	?	?	>	mi	=

Fig. 46 – The initial reality time net (? denotes UNCONSTRAINED)

Furthermore at the beginning all episodes are in `sleeping` and no measurements have been taken yet. Suggesting the first measurement is easy: all episodes are added to candidates in line (1) of SUGGEST, some are also added to follow-ups in (3) but this does not matter at the present moment. Then all but the initial episodes 1 and 4 are pruned away in line (4). Finally episode 1 is dropped in line (6) because episode 4 must be observed both before and after episode 1 so it might be possible to save one observation of 1 if we start with 4. Episode 4 belongs to IN30's history, so we suggest to measure IN30. Let us assume that everything goes well and we measure IN30 = 1. Now MATCH searches for the earliest episode among 4 – 6 with value 1 and finds 4. The episode is in `sleeping`, but there is no other episode that precedes 4, so we go on. There are no previous measurements so that the observation rule is not applied on this turn. But since 4 was observed for the first time, we remove it from `sleeping` and append it to the (previously empty) `open`. Finally, we record that the first measurement m_1 has been associated with episode 4.

Now the second turn starts with a call to SUGGEST. This time all episodes are added to candidates except 4 which is now in `open`. Episode 4 is, however, placed in follow-ups because its relations to 1 – 3 in G have yet to be constrained via the observation rule. In line (4) only episode 1 survives, hence IN29 is suggested. Suppose that again we are successful and measure IN29 = 1. MATCH finds episode 1 and this time we can apply the observation rule to the current measurement m_2 and the previous m_1. $R_{1,4}$ is intersected with UNCONSTRAINED \ {<, m} and via Allen propagation $R_{2,4}$ and $R_{3,4}$ are also updated. The new G is shown in fig. 47. Finally, episode 1 is transferred from `sleeping` to `open` and m_2 is associated with it.

to from	1	2	3	4	5	6
1	=	m	<	o,s,fi,d,=di,f,si,oi,mi,>	?	?
2	mi	=	m	d,f,oi,mi,>	?	?
3	>	mi	=	d,f,oi,mi,>	?	?
4	<,m,o,s,fi,d,=,di,f,si,oi	<,m,o,fi,di	<,m,o,fi,di	=	m	<
5	?	?	?	mi	=	m
6	?	?	?	>	mi	=

Fig. 47 – G after m_1 and m_2

In the next call to SUGGEST episodes 2 – 6 are the initial candidates; 4 is present again, because after m_2 it seems promising to apply the observation rule in the opposite direction. After (4) only 2 and 4 are left, but now deletion rule (6) cuts away 4 on the grounds that we postpone its observation until after 2's because then the two observations of 4 will bracket both 1's and 2's observations and the observation rule can be applied to both at the same time. We suggest IN29 again (but this time because of episode 2). Suppose that we have not waited enough and IN29 is still 1. What happens is exactly the same as on the previous turn except that no changes have to propagated in G. Neither does the sweep-line change and we get the same suggestion again. Now we measure IN29 = 0. MATCH picks episode 2. The observation rule produces the same results that we got the last time via transitivity and thus G does not change. However, 2 moves from `sleeping` to `open` and episode 1 is the first to be shifted from `open` to `closed`.

Let us follow through one more step. The initial candidates now are 3 – 6 with episodes 2, 3 and 6 being follow-ups for the observation rule. Line (4) suffices to remove all candidates except 4 which is now proposed for the second time. If we succeed in measuring IN30 = 1 again, the double application of the observation rule dramatically reduces G (figure 48) whereas the sweep-line remains in place. On the other hand should we measure IN30 = 0, MATCH would associate the observation with episode 5 and we would have missed our last chance to eliminate < and m from $R_{3,1}$ and $R_{3,2}$.

to / from	1	2	3	4	5	6
1	=	m	<	o,s,d	<	<
2	mi	=	m	d,f,oi	<,m,o,fi,di	<,m,o,fi,di
3	>	mi	=	d,f,oi,mi,>	?	?
4	di,si,oi	o,fi,di	<,m,o,fi,di	=	m	<
5	>	d,f,oi,mi,>	?	mi	=	m
6	>	d,f,oi,mi,>	?	>	mi	=

Fig. 48 – G after m_1 – m_4

As we can see from figure 48, the relations between the episodes in the first part of the histories have made good progress in the direction of the corresponding temporal constraints whereas UNCONSTRAINED occurs only between the rearmost episodes.

We leave our example at this point noting that the rest of the matching process will not be very eventful.

3.6.6.2 Second Example

One striking feature of the IN29/IN30 example is the uniqueness of the suggestions: there is always exactly one next measurement which can drive the matching process. This is an artifact of the example; it is different e.g. in cases where the TDS description contains many disjunctive relations. Such a fictitious TDS description is shown in fig. 49; it consists of (identical) histories for three quantities and the only temporal constraint is that the middle episodes must have a pairwise non-empty intersection.

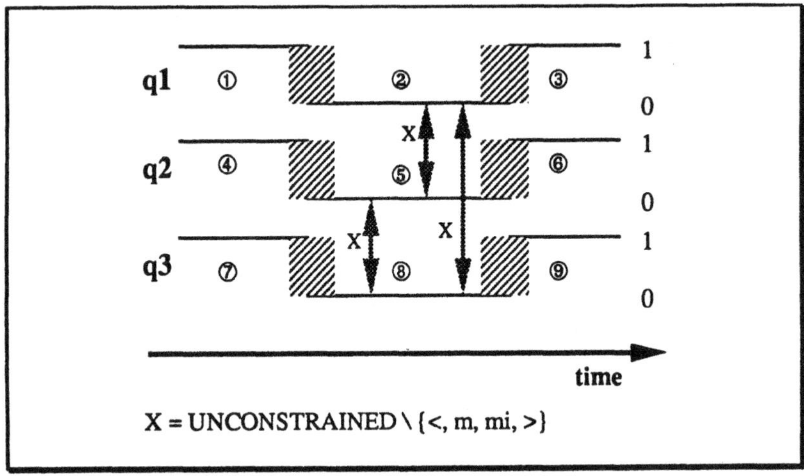

Fig. 49 – A fictitious TDS description

After propagating this constraint and the "meets"-constraints for consecutive episodes we obtain the complete set of temporal constraints shown in fig. 50. Clearly, most of the relations between episodes of different histories are massively disjunctive and the task of planning an economical measurement sequence that matches the TDS description becomes significantly more difficult for an unassisted human observer.

The increased complexity manifests itself in a new feature when we run the temporal matching algorithm on this example. On the first call to SUGGEST all nine episodes are in `sleeping` and therefore initially added to the candidate list. The only deletion rule that is applicable at all in this example is line (4) of SUGGEST which prunes away all episodes except 1, 4, and 7. Consequently the suggestion set contains three elements (all three quantities) and the observer may choose freely which quantity she wants to measure first.

What follows depends on this decision, of course. Let us suppose that the observer chooses q_1 first and measures a 1. In this case the next suggestion again consists of all three quantities because episodes 2, 4, and 7 are to be verified. The situation changes when we observe $q_1 = 0$ next. This observation does not contradict the TDS description, but the next suggestion does not contain q_1 any more. The reason is that a

to from	1	2	3	4	5	6	7	8	9
1	=	m	<	?	<, m, o, s, d	<	?	<, m, o, s, d	<
2	mi	=	m	di, si, oi, mi, >	o, s, fi, d, =, di, f, si, oi	<, m, o, fi, di	di, si, oi, mi, >	o, s, fi, d, =, di, f, si, oi	<, m, o, fi, di
3	>	mi	=	>	d, f, oi, mi, >	?	>	d, f, oi, mi, >	?
4	?	<, m, o, s, d	<	=	m	<	?	<, m, o, s, d	<
5	di, si, oi, mi, >	o, s, fi, d, =, di, f, si, oi	<, m, o, fi, di	mi	=	m	di, si, oi, mi, >	o, s, fi, d, =, di, f, si, oi	<, m, o, fi, di
6	>	d, f, oi, mi, >	?	>	mi	=	>	d, f, oi, mi, >	?
7	?	<, m, o, s, d	<	?	<, m, o, s, d	<	=	m	<
8	di, si, oi, mi, >	o, s, fi, d, =, di, f, si, oi	<, m, o, fi, di	di, si, oi, mi, >	o, s, fi, d, =, di, f, si, oi	<, m, o, fi, di	mi	=	m
9	>	d, f, oi, mi, >	?	>	d, f, oi, mi, >	?	>	mi	=

Fig. 50 – The temporal constraints C in the fictitious TDS description

second observation of 2 would not serve any purpose right now and an observation of 3 would lead to a contradiction (no measurement may be associated with 3 as long as 1 and 4 are still `sleeping`). Therefore the matching algorithm insists on finding corroborations for episodes 4 and 7; only after this has been accomplished is q_1 again a member of the suggested set (for a second application of the observation rule).

The example illustrates how the suggestion of further measurements depends critically on the non-deterministic choices made before. In principle we could precompute all possible sequences of choices and corresponding suggestion sets. The number of sequences is, however, quite large and in view of the moderate execution times of SUGGEST we resolve the tradeoff between space and time in favor of time, i.e. computing the suggestions on the fly.

3.7 Quantitative Temporal Information

Everything we have said in this chapter about the representation and matching of TDSs and about the development of the temporal matching algorithm has been based on the universal premise that the temporal constraints in TDS descriptions are of a purely qualitative nature. This assumption was made more because it was convenient than well-founded and because it enabled us to discuss the key concepts in a particularly concise and compact form. But now we take a second look at it: are qualitative temporal relations all there is? When we acquired the diagnostic knowledge for the MOLTKE project it did not take us long to discover TDSs where qualitative constraints are complemented by quantitative constraints.

A fairly prototypic example is given in fig. 51 which occurs in several variations in MOLTKE's domain. The tool arm (and other arms) of the CNC center can move along all three spatial axes. For each movement the control computes a time interval before the end of which the movement should be completed under normal operating conditions. If the time is exceeded, the control reports a mechanical fault. In practice, however, the time-out alarm can be caused by a range of different conditions – with different repair actions. Apart from the expected mechanical cause – movement has been blocked by a collision or by turnings stuck in the transmission – the movement may have been too slow, or even the alarm mechanism itself may be faulty. These different causes can be distinguished only on the basis of TDSs. The "slow movement" case is particularly confusing, because the time-out is reported and nevertheless the arm reaches its target position; if one sees only the final state, this is difficult to explain and the alarm mechanism is likely to be suspected. What really happens in this case, is depicted in the following diagram:

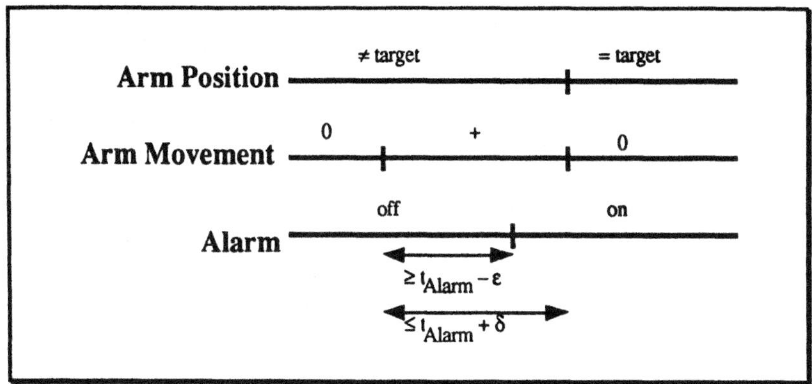

Fig. 51 – The time-out alarm TDS description

Here t_{alarm} denotes the time from the beginning of the movement to the beginning of the alarm, ε is the maximum tolerance for an early alarm and δ is the maximum delay which may be attributed to low speed. As we can see clearly in the diagram, the qualitative relations between the episodes are augmented by duration constraints which here take the form of absolute upper or lower bounds on the distance between two episode end points.

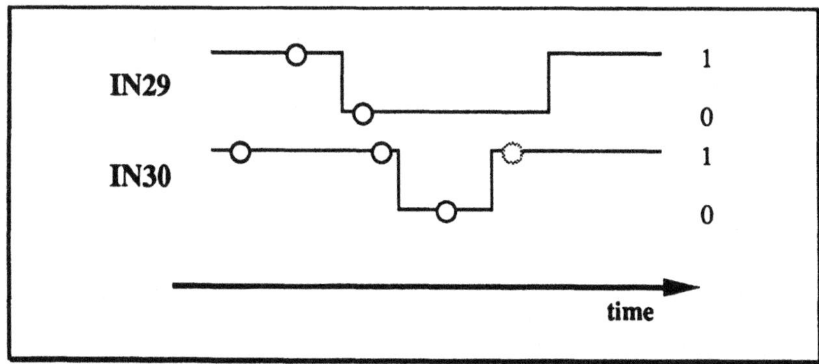

Fig. 52 – A looping situation for the temporal matching algorithm

Rather surprisingly, duration constraints can not only help in specifying TDS descriptions with quantitative information; they can also be used to control the matching of otherwise purely qualitative TDS descriptions. We can illustrate this with the help of our former example. Imagine that we have successfully made the observations shown by solid circles in fig. 52. Our algorithm suggests that we measure IN30 again and tells us to expect a 1 (dotted circle). But suppose that in reality both IN29 and IN30 remain stuck at 0 instead returning to 1. In this situation the matching algorithm will cycle forever, suggesting to measure IN30, expecting IN30 - 1 and accepting IN30 = 0 instead because it believes that the transition will eventually happen. This behavior is perfectly justified as long as there is no information available about the length of IN30's middle episode. Things change when we know in addition that e.g. IN30 = 0 for at most 5 seconds. We can then abort the matching algorithm and report failure as soon we have failed to measure IN30 = 1 for five seconds after the first IN30 = 0 observation.

The two examples show that quantitative information is essential in realistic domains and that therefore we must extend our representation formalism and matching techniques appropriately. The question is how far should we go? [Schmiedel88] discusses a hierarchy of quantitative temporal constraints of varying expressive power and demonstrates how the form of the constraints directly influences the complexity of the inference mechanisms. He also shows that already for seemingly simple constraint languages the global consistency and minimal labeling problems become computationally intractable. The results are borne out by experiences of Kautz [Kautz88] during implementation of the duration reasoner sketched in [Allen83]: despite sacrifices in performance it turned out to be nearly impossible to make the reasoner both sound and complete. For these reasons we have adopted a pragmatic standpoint: we augment our basic framework by absolute upper and lower bounds on durations between episode end points, since this is sufficient for all examples that we encountered. There is no theoretical reason for doing so and one can certainly construct more complex TDSs but already the next steps in the hierarchy of [Schmiedel88] are computationally so much more expensive that it is doubtful whether a system using them would be of much practical use.

In the following three sections we describe how TDS descriptions, definition 20 and the matching algorithm are affected by the extension. The description will be incremental, so that the reader will want to refer back to section 3.6 for complete information about the basic framework. The general philosophy of the extension will be to treat qualitative and quantitative information about a TDS as two different bodies of knowledge. Developing a combined inference engine for both qualitative and quantitative constraints is still an open problem and we prefer a more robust solution. For this we pay the price of sometimes making less inferences about reality episodes than possible and thus needing extra measurements before we can be certain about the result of the match.

3.7.1 TDS Descriptions with Durations

Extending the form of TDS descriptions is relatively straightforward. In addition to the qualitative constraints in the TDS time net we have a set of absolute (i.e. real-valued) bounds on duration between interval end points. By LOWER (A, B) we denote the lower bound on the duration between the starting point of A-interval A and the end point of A-interval B. UPPER (A, B) is the corresponding upper bound. We also allow LOWER (A, B) = $-\infty$ and UPPER (A, B) = ∞ denoting the absence of any information about the lower (upper) bound. Note that A and B may or may not be identical.

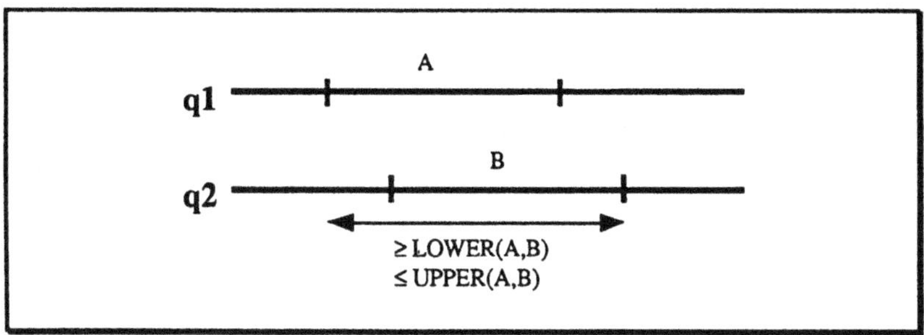

Fig. 53 – UPPER and LOWER duration bounds

Without going into detail we can define time nets with duration bounds as four-tuples ⟨V, C, UPPER, LOWER⟩ and state their semantics in much the same way as we did for ordinary time nets in section 3.2.3. If for example LOWER(A,B) = c_1 and UPPER(A,B) = c_2 then

$$[\![\langle\{A,B\}, C, \text{UPPER}, \text{LOWER}\rangle]\!] := \{\ D \in [\![\langle\{A,B\}, C\rangle]\!] \mid D(A) = [A^-; A^+],\ D(B) = [B^-; B^+],\ c_1 \le B^+ - A^- \le c_2\ \}.$$

The semantics definition shows that some Allen relations are special cases of duration bounds. The reasons for keeping both are twofold: not all primitive Allen relations are duration constraints and the inference rules embodied in the basic matching algorithm and in Allen's path-consistency algorithm are stronger than those for general duration constraints.

We obtain the following new definition for TDS descriptions:

DEFINITION 21: A *TDS description with duration bounds* is a 5-tuple
⟨Q, H, C, LOWER, UPPER⟩ where
• Q, H, C are as before and
• LOWER and UPPER are functions

$$\text{LOWER, UPPER: EPI} \times \text{EPI} \rightarrow \mathbf{R} \cup \{-\infty, \infty\}$$

where EPI is the set of all episodes in H and
$\forall$E, E' $\in$ EPI: LOWER(E, E') $\leq$ UPPER(E, E').

The associated TDS time net with duration bounds is defined in the obvious way.

All definitions relating to TDS instances remain unchanged except that now we speak about models of the TDS time net *with* duration bounds.

3.7.2 Matching with Durations - The Definition

The key question in defining a match between a measurement sequence and a TDS description with duration constraints is under which conditions measurements can determine durations. Consider the situation in fig. 54 where m_1 is the last observation associated with A, m_2 is the first associated with B, m_3 the last for C, and m_4 the first for D.

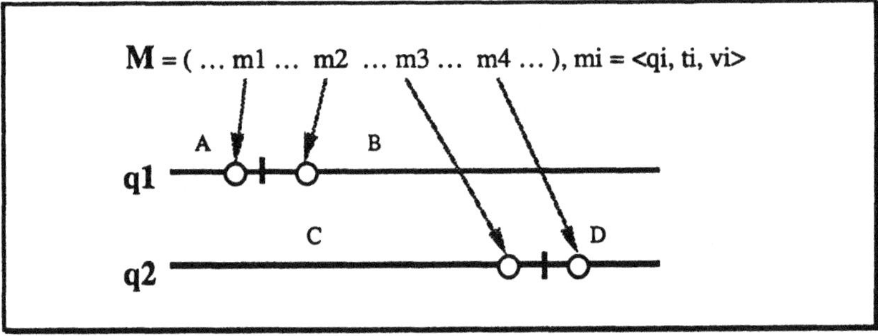

Fig. 54 – Relevant measurements for verifying duration constraints

Obviously we cannot determine the duration between B^- and C^+ *exactly* given only discrete measurements, but we can approximate it from above and below by taking differences of t_i:

$$t_3 - t_2 \le C^+ - B^- \le t_4 - t_1$$[31].

We can immediately transform this inequality into sufficient conditions for B and C satisfying their lower and upper bound constraints:

$$t_3 - t_2 \ge \text{LOWER}(B, C) \Rightarrow C^+ - B^- \ge \text{LOWER}(B, C)$$

$$t_4 - t_1 \le \text{UPPER}(B, C) \Rightarrow C^+ - B^- \le \text{UPPER}(B, C)$$

The second condition can be strengthened somewhat, because instead of taking the last measurement for the predecessor episode in the same history we can take the last measurement for *any* episode preceding B, and the dual applies to C. The two rules which we call the lower and upper bound rules directly supply the additional conditions for "M matches S":

DEFINITION 22: A measurement sequence $M=\{\langle q_i,t_i,v_i\rangle\}_{i=1,\dots,n}$ *matches_with_durations* a TDS description with duration bounds $S=\langle Q,H,C,\text{LOWER},\text{UPPER}\rangle$, iff there is an instance $\langle Q,H,D\rangle$ of S that satisfies the following conditions:

(i) – (v) as in definition 20

(vi) $\forall\, q, q' \in Q; \forall\, E = \langle I,v\rangle \in H_q, E' = \langle I',v'\rangle \in H_{q'}$: $\text{LOWER}(E,E') = c > -\infty$
$\Rightarrow \exists\, \langle q, t, v\rangle, \langle q', t', v'\rangle \in M$: $I^- \le t' < t \le I'^+ \wedge t' - t \ge c$
(if the lower bound is non-trivial, there are two observations that allow application of the lower bound rule)

(vii) $\forall\, E'', E'''$ episodes in H: $\text{UPPER}(E'',E''') = c < \infty$
$\Rightarrow \exists\, E = \langle I,v\rangle \in H_q, E' = \langle I',v'\rangle \in H_{q'}$: $\exists\, \langle q, t, v\rangle, \langle q', t', v'\rangle \in M$:
$C_{E,E''} \subseteq \{<, m\} \wedge C_{E''',E'} \subseteq \{<, m\} \wedge t \le I^+ < I'^- \le t' \wedge t' - t \le c$
(if the upper bound is non-trivial, there are two observations that allow application of the upper bound rule)

This definition has an important implication for the kinds of duration constraints we can hope to verify. Because of the way that (vi) and (vii) interact, it is principally impossible for us to satisfy both when LOWER(E,E') = UPPER(E,E'), i.e. when we

---

[31] There is a strong resemblance between this approximation and the representation formalism for incomplete knowledge, called *rough sets* [Pawlak82]. Indeed, we could rephrase the inequality in the following way: instead of knowing the exact set $D = \{x \mid x \le C^+ - B^-\}$ we only know its rough set approximation $<\underline{D} = \{x \mid x \ge t_3 - t_2\};\ \overline{D} = \{x \mid x \le t_4 - t_1\}>$.

specify an exact duration between two episode end points. Just as in the case of densely spaced transitions discussed in section 3.5 the nature of our measurements puts a limit to what can (or should) be expressed in a TDS description. If the minimal granularity of the measurement sequence is ε, then all duration constraints in the TDS description should satisfy UPPER(E,E') – LOWER(E,E') $\geq 2\varepsilon$, or we are certain to get a failure from insufficient data.

We now verify that our extension preserves the weak soundness and weak completeness properties of definition 20.

> THEOREM 6' (weak completeness):
> $\forall$ measurement sequences M: $\forall$ TDS descriptions with durations S:
> M weakly determines S $\Rightarrow$ M matches_with_durations S.

PROOF: We use the same line of argument as before in the proof of theorem 6. Because of our separate treatment of qualitative and quantitative constraints cases (i) – (v) still go through. We thus have to show

$(\forall$ TI instance of S: $\neg$ (vi)) $\Rightarrow$ X and

$(\forall$ TI instance of S: $\neg$ (vii)) $\Rightarrow$ X.

Case (vi):

$\forall$ TI instance of S: $\neg$ (vii) $\Rightarrow$

$\exists\, q, q' \in Q; \exists\, E = \langle I,v\rangle \in H_q,\ E' = \langle I',v'\rangle \in H_{q'}$:

LOWER(E,E') = $c > -\infty$

$\wedge \neg\, [\, \exists \langle q, t, v\rangle, \langle q', t', v'\rangle \in M$: $I^- \leq t' < t \leq I'^+ \wedge t' - t \geq c\,]$.

Take an arbitrary instance TI of S with BASIS(TI) = Int(M) and gran(TI) $\geq 2 \cdot$ gran(M). Move I^- towards the time of the first observation of I and I'^+ towards the time of the last observation of I' until $I'^+ - I^- < c$. The new instance TI' is no longer an instance of S, but M is compatible with TI'. Without loss of generality we can assume that we can move I'^+ and I^- in such a way that gran(TI') $\geq 2 \cdot$ gran(M). But if TI' occurs in Int(M), then no instance of S occurs in Int(M).

Case (vii):

$\forall$ TI instance of S: $\neg$ (vii) $\Rightarrow$

$\exists\, q, q' \in Q; \exists\, E = \langle I,v\rangle \in H_q,\ E' = \langle I',v'\rangle \in H_{q'}$:

UPPER(E'',E''') = $c < \infty$

$\wedge \neg\, [\, \exists\, E = \langle I,v\rangle \in H_q,\ E' = \langle I',v'\rangle \in H_{q'}$: $\exists \langle q, t, v\rangle, \langle q', t', v'\rangle \in M$:

$C_{E,E''} \subseteq \{<, m\} \wedge C_{E''',E'} \subseteq \{<, m\} \wedge t \leq I^+ < I'^- \leq t' \wedge t' - t \leq c\,]$.

Construct TI' as in case (vi), but move I^+ and I'^- further *apart* until $I'^- - I^+ > c$. ■

THEOREM 7' (weak soundness):

$\forall$ measurement sequences M: $\forall$ TDS descriptions with durations S:

M matches_with_durations S $\Rightarrow$ M weakly determines S.

PROOF: All of the proof of theorem 7 carries over. It remains to be shown that the reality-episodes in Int(M) form a model not only of the qualitative constraints, but also of the complete TDS time net with the duration bounds. But this follows directly from our derivation of the lower and upper bound rules at the beginning of the section. ■

3.7.3 A Matching Algorithm for TDSs with Duration Constraints

The discussion of duration in the last two sections benefited from the strict separation between the basic definitions which in principle remained unchanged and the extensions for quantitative constraints. The separation could even be maintained in the new definition 22 which acts as the specification for an extended matching algorithm. Here at last we are forced to abandon this form of treatment: at *each* step of the algorithm we want to use *both* qualitative and quantitative elements of the TDS description to their fullest extent, and this precludes any kind of sequential approach. We will therefore examine each of the algorithms from section 3.6 in turn deciding which parts to keep and which modifications and extensions to make.

3.7.3.1 MAIN

The logic of the main loop is not affected in any way by the introduction of duration constraints. It is not even necessary to introduce new temporary data structures for keeping record of the partial information about the durations of reality-episodes. Definition 22 shows that the only additional data we need are the exact time points of the observations. We can recover from the association list which contains all available data about the measurements made up to the current moment[32].

---

[32] For faster access one might nevertheless want to organize storage of observation times differently in a concrete implementation of the algorithm.

3.7.3.2 SUGGEST

Recall that suggesting the next observation happens in two different phases: we first compute a superset of the suggestions which is afterwards pruned according to the deletion rules. Consider the generation first: the two operations in this phase detect opportunities to push the sweep-line forward and to apply the observation rule, respectively. What is the analogous operation for determining suitable observations for the upper and lower bound rules? According to condition (vi) of the matching it takes two measurements (one for E, one for E') to verify a LOWER(E, E') constraint. From the inequality given in condition (vi) of definition 22 it is also clear that the best choice for the two observations are the first observation of E and the last observation of E'. For them the approximation error is minimal and thus the chance of their satisfying the inequality is maximal. Observing an episode at the earliest possible opportunity is already taken care of: if an episode has not been associated with a measurement before, it is still a member of `sleeping` and therefore automatically added to candidates. However, if an episode has already been moved to `open` it might be overlooked on subsequent calls to SUGGEST, if it does not accidentally take part in an application of the observation rule. This is especially undesired for episodes E with a non-trivial LOWER(E,E), because these definitely need to be observed twice. Unfortunately there is no way of guessing when a reality-episode will end[33]. We therefore suggest to look for an episode E whenever for some E' there is a lower bound LOWER(E',E) left that has not been verified by observations. We do this by defining for each time point t ∈ Int(M) two auxiliary functions

$$\text{firstObs}_t, \text{lastObs}_t\text{: } (\texttt{open}_t \cup \texttt{closed}_t) \to \mathbb{R}$$

where $\texttt{open}_t$ and $\texttt{closed}_t$ are the values of the `open` and `closed` lists of the basic matching algorithm at time point t. For each episode in $\texttt{open}_t \cup \texttt{closed}_t$ the functions firstObs_t and lastObs_t give the times of the first (last) observation associated (so far) with the episode. The functions can be thought of as look-up functions into the association list M. We can then add the following statement to SUGGEST after line (1):

```
(1')  For each episode E ∈ open do:
      if E' ∈ open ∪ closed and LOWER(E', E) > -∞ and
      last Obs(E) - first Obs(E') < LOWER(E', E)
            then candidates ← candidates ∪ {E}.
```

---

[33] except in the presence of further external knowledge.

Upper bounds are different: verifying UPPER(E',E) takes two observations, too, but not of E and E'. In particular it is never necessary to observe an episode twice as in the case of lower bounds. Since SUGGEST is guaranteed to propose the observation of each episode at least once, we do not have to take a special action for upper bound constraints. There are no new deletion rules relating to the duration constraints. For episodes which have been added to candidates in line (1') we could, however, include in the suggestion the earliest observation time that would satisfy condition (vi) of definition 22; if LOWER(E', E) >> 0, the person performing the measurements could turn to other tasks and resume the matching process at a later time when the lower bound has definitely been exceeded[34]. But how do the existing deletion rules interact with (1')? All of these have the common motivation that we can safely drop an episode from the present candidate list, if we must observe it again later anyway. By the same argument we can see that the deletion rules do not interfere with (1'): no deletion rule will ever cancel the last observation for an episode - observations are always only postponed - and for condition (vi) we are only interested in the last observation. So the pruning phase of SUGGEST is not changed.

3.7.3.3 MATCH

Violations of both the lower and upper bound constraints are detected and reported in MATCH. When we use the term "violation" we have to be aware of the fact that conditions (vi) and (vii) in definition 22 are sufficient but not necessary: failure to meet any of these conditions therefore only implies termination with "insufficient data". We can obtain necessary conditions by taking the duals of (vi) and (vii), but in both cases the "insufficient data" condition is detected. Since we prefer to abort the matching process as early as possible so that effort is saved and matching can be restarted for a second try, we stick to the sufficient conditions. Conflicts with condition (vi) can be detected immediately after choosing an episode to be associated with the latest measurement. We add statement (2') after line (2) of MATCH:

```
(2')  For all episodes E' ∈ open, E'' ∈ (open ∪ closed) do:
          if lastObs(E') - firstObs(E'') < LOWER(E'',E')
          and (C_E',E ≤ {<, m} or R_E',E ≤ {<, m})
              then report insufficient data, stop.
```

[34] unless, of course, intermediate measurements are desired to keep the granularity low.

The intention behind (2') is to catch cases where a second observation of E' is needed to verify LOWER(E", E'), but this is no longer possible now that E has been observed.

UPPER(E, E') constraints are treated similarly. As soon as the difference between the current moment (given by the time of the latest observation) and the time of the last measurement before E exceeds UPPER(E, E') we know for sure that any observation after E' will definitely come too late. We insert line (2") after (2'):

```
(2'') For all episodes E' ∈ open ∪ {E}, E'' ∈ open ∪ closed:
      if for all episodes E''' with C_E''', E'' ≤ {<, m}:
      t - lastObs(E''') > UPPER(E'', E')
      then report insufficient data, stop.
```

3.7.4 Evaluation

The advantages and disadvantages of using the extension of the basic matching algorithm to duration bounds can be summarized as follows.

Firstly, the added expressive power of the TDS description language allows the specification of TDSs in which qualitative and quantitative elements are mixed. The time-out alert situation is an example from MOLTKE's domain. Secondly, we have shown that even for inherently qualitative TDSs it can pay to specify available quantitative information in the TDS description, because the discriminatory power and hence the performance of the matching algorithm is enhanced.

Clearly, the extension covers only some of the conceivable forms of quantitative temporal information. While its expressiveness is sufficient for the applications we studied, this is not guaranteed in different domains. In fact, studying the utility of our approach in new domains is one of the obvious directions for future work. Nevertheless, the decision to restrict the quantitative part of the language to absolute duration bounds has its benefits, too: despite the extension the time complexity of the augmented matching algorithm is of the same order as that of the basic algorithm. Admittedly, this result depends on the weak coupling between qualitative and quantitative information in the matching algorithm; developing an efficient algorithm for determining consistency in a time net with duration bounds, which uses both sources of information simultaneously, is still an open problem.

4 Incorporating TDSs into Existing Diagnostic Paradigms

The presentation in chapter 3 has focused primarily on the technical aspects of temporal matching as an end in itself. Although our example TDSs were drawn from a diagnostic domain, we have kept task-specific discussions to a minimum.

The principal reason for doing this was our goal to characterize temporal matching in as general a way as possible so that its role as a universal building block for diagnostic expert systems[35] would become apparent. Choosing sides too early – i.e. describing our technique with respect to only one of the competing diagnostic paradigms – would have obscured its generality.

We now come back to our original motivation and demonstrate in this chapter how temporal matching can be useful in the two dominant paradigms: associative/heuristic and model-based diagnosis. We base our discussion in both cases on existing diagnostic systems which are fairly prototypic of the two paradigms. Furthermore the representative we have chosen for model-based diagnosis (GDE/SHERLOCK) has set a certain standard regarding terminology and customarily new extensions are characterized in relation to its framework. Associative diagnosis is represented by the MOLTKE system, for which temporal matching was developed originally.

---

[35] or even for different tasks (see the evaluation in section 5.3).

4.1 Associative/Heuristic Diagnosis: MOLTKE

4.1.1 MOLTKE[36,37]

The MOLTKE expert system [Althoff et al. 88] was developed at the university of Kaiserslautern as a platform for studying various advanced diagnostic techniques within the traditional framework of associative diagnosis [Althoff et al. 89a], [Althoff et al. 89b], [Rehbold89], [Nökel89a], [Nökel89b]. Each extension was motivated by the special requirements of MOLTKE's application domain, the diagnosis of a CNC[38] machining center.

In chapter 3 we have already seen examples of TDSs from MOLTKE's domain. Here we explain how their treatment is actually incorporated into MOLTKE's otherwise static framework. We start by giving an overview of MOLTKE's basic terminology and architecture paying special attention on how time-independent symptoms are handled. Section 4.1.2 deals with the integration of TDSs into MOLTKE's knowledge base and with the interface between the temporal matching algorithm and the diagnostic procedure. We emphasize that our extension preserves the general philosophy of MOLTKE and is effective in the sense that it does not lead to computational overhead when time-independent symptoms are processed.

A detailed description of MOLTKE's architecture can be found in [Kockskämper89]; we will instead introduce the basic terminology in an informal way and take a closer look only at the mechanisms by which MOLTKE obtains information about symptoms.

In line with the approach taken in most associative diagnostic systems ([Richter89], [Puppe87]), the basic concepts in the ontology are symptoms, tests and diagnoses. *Symptoms* are observable quantities which take on different values depending on the machine state. *Tests* are procedures which determine the value of one or more symptoms. There may be more than one test for a given symptom, with tests differing in precision, cost, and a number of other criteria. For the majority of symptoms,

36 Models, Learning and Temporal Knowledge in an Expert System for Technical Diagnosis

37 Throughout this section we refer to version 2.0 of the MOLTKE system.

38 Computerized Numerical Control

however, there is a 1-1 correspondence between a symptom and a test which measures exactly the value of this one symptom. *Diagnoses* are characterized by formulae over symptom values. An important modularization aspect in MOLTKE's knowledge base is that common parts in the characteristic formulae of different diagnoses are factored out and the diagnoses sharing such a part are grouped together in a *context* which has the common part as its precondition. Iterating this step leads to the construction of a context heterarchy which is closely related but not identical to the component structure and to the functional structure of the machine.

The diagnostic procedure consists of establishing the truth of the characteristic formulae along a path from the root of the context heterarchy down to a final diagnosis. The choice of the next symptom to be observed is made on the basis of expected information gain, i.e. the status of the preconditions of competing contexts are examined. In addition, heuristic expert knowledge may be specified explicitly in the form of rules which override the default strategy. In both cases, MOLTKE's probing strategy conforms to the *parsimony principle:* symptom values are determined *at most once*; due to the static view of the machine symptom values never change so that one observation is always sufficient. As we will see, for TDSs it is almost *necessary* to violate the parsimony principle.

The interaction of all these entities can best be demonstrated by a generic example. Let us assume for this purpose that the system is trying to figure out which test it should carry out next. In particular it is trying to apply the rule in fig. 55 thereby establishing the value of the symptom `IOStateOut-20`[39]. What happens?

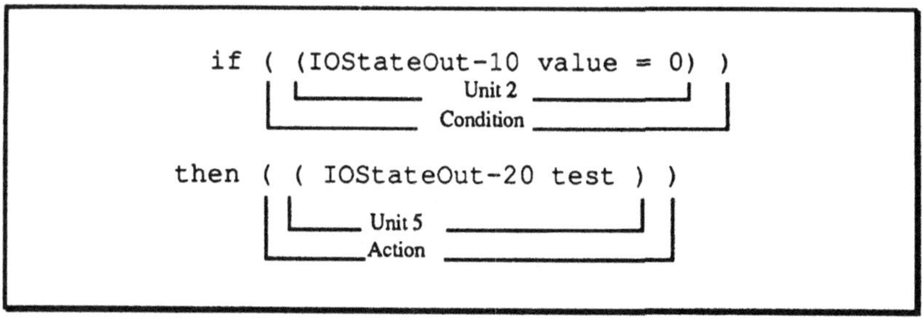

```
   if ( (IOStateOut-10 value = 0) )
        |L_________ Unit 2 __________J |
        L__________ Condition _________J

then ( ( IOStateOut-20 test ) )
       | L____ Unit 5 _________J |
       L______ Action _________J
```

Fig. 55 – Example rule for testing a time-independent symptom

[39] None of these identifiers is significant for understanding the example.

Fig. 56 describes the internal representation of the example rule from fig. 55. We will give a trace through fig. 56 assuming that the system is currently in context `Tool-Magazine`, because its precondition ① has been evaluated to true.

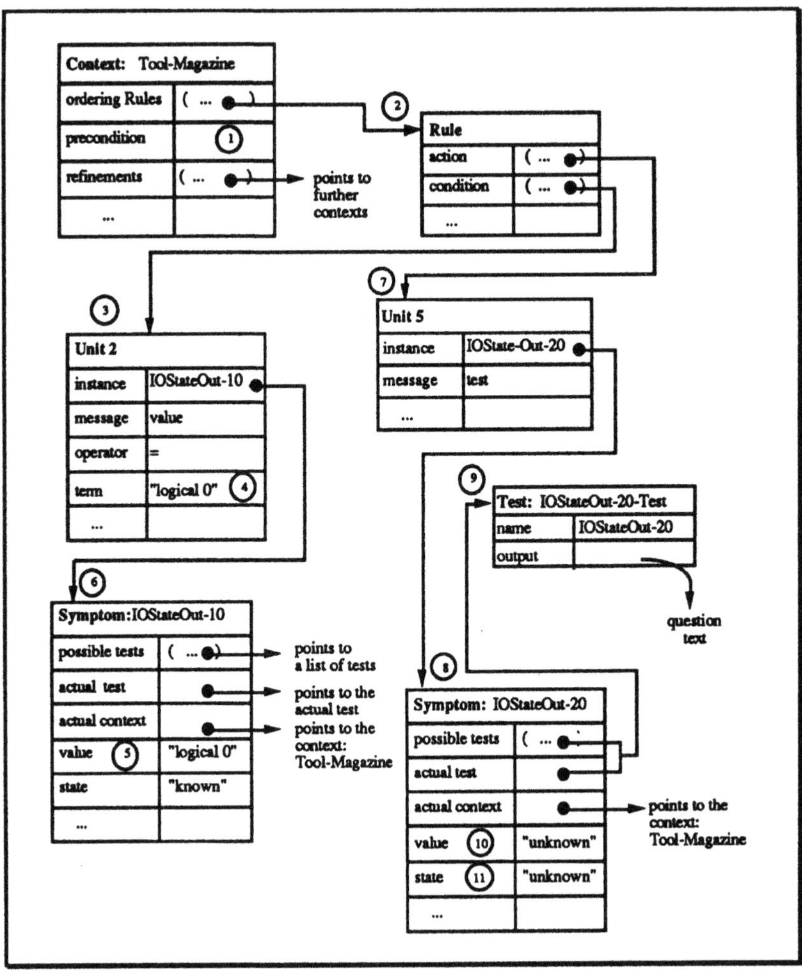

Fig. 56 – Internal representation of the example rule in fig. 55

Each rule ② consists of a condition and an action depending on that condition. Both condition and action parts are made up of literals which are called *units* in MOLTKE. There are different types of units for various types of tests and actions: in our example a unit ③ of type `Unit2` (technically, an instance of the class `Unit2`) compares a term ④ to the value of the symptom ⑥ which belongs to this instance of `Unit2`. Each unit class has its own matching algorithm which is called to perform the test or action whenever one of its instances is evaluated. As our example shows, state = known flags that the symptom has been observed earlier in the session; therefore its value can simply be read from the value slot ⑤, the match between this instance of Unit2 ③ and the symptom ⑥ is successful and the condition of the rule is satisfied. Now we can perform the action which is connected to an instance of Unit5 ⑦. The matcher of the Unit5 class sends the message test to the symptom IOStateOut-20 ⑧. In contrast to symptom IOStateOut-10, this symptom's state is unknown, so that a test has to be performed. Suppose that IOStateOut-20-Test ⑨ is chosen. Evaluating a test for a time-independent symptom results in an appropriate question being put to the user and reading the observed value into the value slot of the symptom ⑩. Afterwards the state slot ⑪ is set to known ensuring that the symptom value will never be determined a second time. Following the test other rules of the context Tool-Magazine are applied to choose the next observation. If none are left, context preconditions are reevaluated using the new symptom values (including the value of ⑩) to select the next context.

It is important to note that the sequence of selecting contexts and asking questions, which seems completely under the control of the diagnostic system, in fact serves only as a guiding-line for the technician. Early in the knowledge acquisition process it turned out to be critical for acceptance that the user can deviate from the dialog whenever he wants to. We achieve this flexibility by letting the user decide at each test whether he wants to answer the question or whether he wants to switch to other contexts and evaluate other rules. In that case, the slots state and value of the symptom whose test was suspended remain unknown.

4.1.2 Integrating TDSs into MOLTKE

Although the static view of the machine is a useful abstraction, there are certain symptoms (TDSs) in MOLTKE's domain which cannot be meaningfully interpreted under this assumption. Two examples have been given in chapter 3; other examples deal e.g. with feedback effects in the drive system. Our discussion of temporal matching has shown that the detection of TDS occurrences requires more than one measurement of the same quantity and that the temporal order of the measurements is significant. Both properties are contrary to MOLTKE's parsimony principle. In the basic system it is not even possible to translate either of our running examples into a characteristic formula, because there is no way in the rule language to temporally index literals.

For these reasons we have to extend the standard treatment for symptoms to cover TDSs. The task is twofold:

- We have to extend the representation language for characteristic formulae of TDSs.
- Correspondingly, we have to design a new type of test which can be used to establish the "value" of a TDS.

4.1.2.1 The Representation of TDSs in the Knowledge Base

At the beginning of section 3.2 we pointed out the dilemma of designing a uniform description language for TDSs. We concentrated at that point on the requirements of temporal matching and proposed TDS descriptions as the internal representation.

In designing an external syntax for TDSs other factors must be emphasized:

- *Readability:* For the purposes of documentation and debugging the syntax should be close to the format used by human experts for describing TDSs.
- *Efficiency:* The syntax should be flexible towards specification of redundant information. In general, where information about one part of the TDS description can be deduced from the specification of another, the system should be able to do so. If the user prefers to specify both parts explicitly, the deductive power should be used to check the user input for consistency.
- *Relation to internal representation:* Since the external syntax has to be translated into a TDS description, it must be possible to parse it unambiguously.

The syntax that we are currently using satisfies the last two requirements to a high degree. By following the general syntax for units we not only achieve uniformity with the rest of the knowledge base syntax, but even the actual parser software can be reused without change.

Fig. 57 shows the specification of the familiar IN29/IN30 TDS in this syntax. The specification consists of three parts:

- In part A all episodes are declared and given a symbolic name for reference in parts B and C.
- Part B contains the declarations of the qualitative temporal constraints.

- Part C (which is empty in the example) contains the declarations of the duration constraints.

```
((TNET
        (IN29 = 1 E1) ┐
        (IN29 = 0 E2) │
        (IN29 = 1 E3) │
        (IN30 = 1 E4) ├      A
        (IN30 = 0 E5) │
        (IN30 = 1 E6) ┘
        (E1 (m) E2) ┐
        (E2 (m) E3) │
        (E4 (m) E5) ├        B
        (E5 (m) E6) │
        (E2 (oi) E4)│
        (E2 (o) E6) ┘
))
```

Fig. 57 – Specification of the IN29/IN30 TDS in the external syntax

Notice that part B need not be specified completely; in the example it is sufficient to declare which episodes form a history and how E2, E4 and E6 overlap. During compilation the relations from part B are added to an initially unconstrained time net and their consequences are computed using Allen's propagation algorithm. Inconsistencies are detected easily when the propagation produces an empty relation. Also at this stage all relations are checked for the convexity property.

Admittedly, the readability of our syntax (especially for debugging) is at best medium. It appears that the only real improvement in this respect would be a radical move to a graphical interface. Our experiences with graphic representations of time nets with disjunctive relations indicate, however, that this would be a highly non-trivial and probably computationally expensive step and we reserve it as future work.

A BNF definition of the external syntax can be found in appendix A.

4.1.2.2 Temporal Matching as a Special Form of Test

The fundamental design decision is on which level of the architecture time-independent symptoms and TDSs should be treated alike so that no modifications are needed from there upward. There is more than one possible answer to this question and the decision has to be made on the basis of both conceptual and pragmatic (software engineering) arguments. On the one hand, pragmatic reasons tend to push the dividing line downward so that as little as possible of the architecture needs to be duplicated; for reasons of conceptual clarity, on the other hand, it is often desirable to make the distinction on a higher level. After having analyzed the advantages and disadvantages of different scenarios we decided to use the same formalism down to and including the level of symptoms. All time-specific processing is encapsulated on the next lower level, in the form of a special new type of test. Since time-independent symptoms still use the same matcher as before – i.e. we do not use the temporal matcher on a trivial TDS –, the overhead of temporal matching occurs only in the cases where it is needed.

This decision has important implications for the semantics of TDSs in MOLTKE. While the observation(s) needed to establish the value of a symptom are carried out under the responsibility of the test, the parsimony principle is a built-in property of symptoms. This means that although it is possible to have tests which execute more than one measurement, the TDS *as a whole* is still only observed at most once. At first sight this seems illogical, because we might infer the absence of a TDS from a single ill-timed attempt at observing it which happens to fail. The rule language, however, already permits to express strategies for retrying observations in a more flexible and individual manner than could be easily built into a general "TDS manager". The one remaining important consequence of the parsimony principle concerns negation: since we interpret a reference to a TDS description in a characteristic formula in the sense of "an instance of the TDS has occurred during some subinterval of the whole diagnostic session", a negated reference would have to be interpreted as the "absence of the TDS throughout the diagnostic session." This cannot in general be guaranteed using only the discrete measurements available; hence we forbid negation of TDSs in rules and context preconditions.

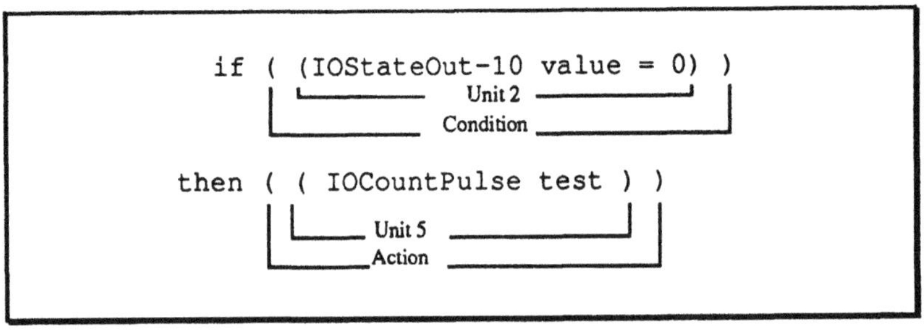

```
if ( (IOStateOut-10 value = 0) )
      Unit 2
     Condition

then ( ( IOCountPulse test ) )
      Unit 5
      Action
```

Fig. 58 – Example rule referring to a TDS

We are now ready to describe the treatment of TDSs in analogy to section 4.1.1. Rules which trigger the observation of TDSs are written in exactly the same way as in the time-independent case. An example is shown in fig. 58. This example uses the same assumptions as the earlier example in fig. 55. The only difference is that here `IOCountPulse` is the name of a TDS.

Again fig. 59 shows the internal representation of the example rule in fig. 58.

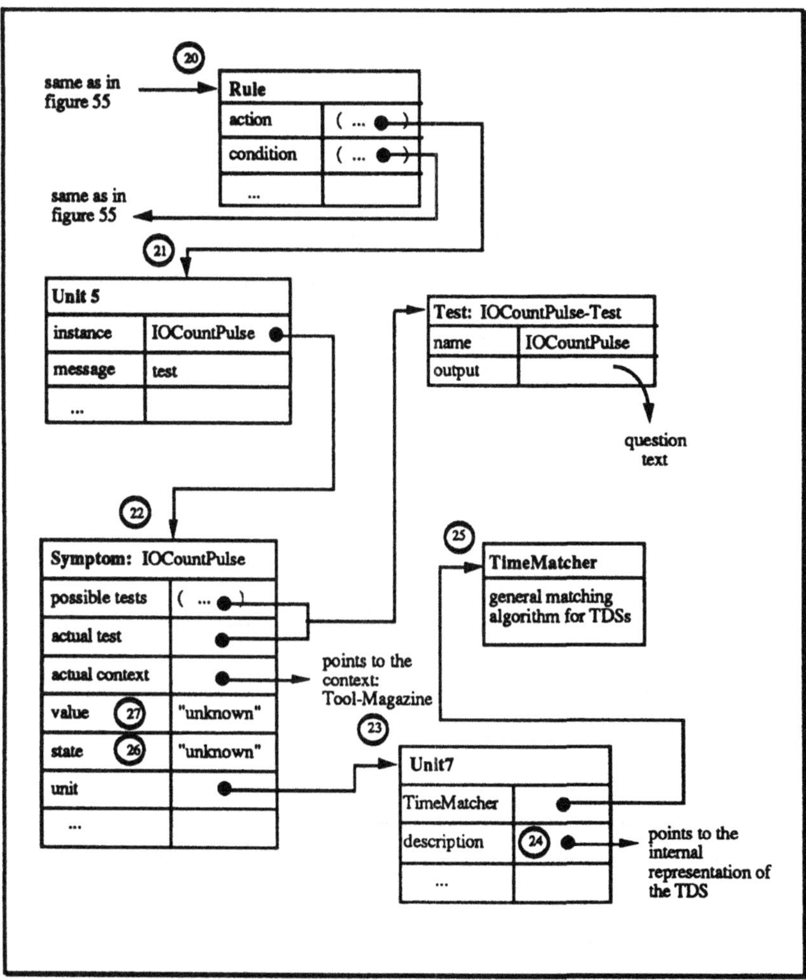

Fig. 59 – Internal representation of the example rule in fig. 58

Assume that the condition of the rule ⑳ is evaluated to `true` in the same manner as in section 4.1.1. This enables the instance of `Unit5` ㉑ connected to the `action` slot to be evaluated. The matcher of the class `Unit5` sends the message `test` to the TDS `IOCountPulse` ㉒. As compared to a time-independent symptom, the `state` slot of a TDS can take on a third value (`in_progress`), which is explained below, in addition to `known` and `unknown`. Also a slot `unit` appears, which points to an instance of the new class `Unit7` ㉓. Units of type `Unit7` contain the internal representation of a TDS ㉔ in a form that corresponds to definitions 10 and 11. Just as classes of ordinary units possess a matching algorithm, the class `Unit7` is connected to the temporal matcher ㉕.

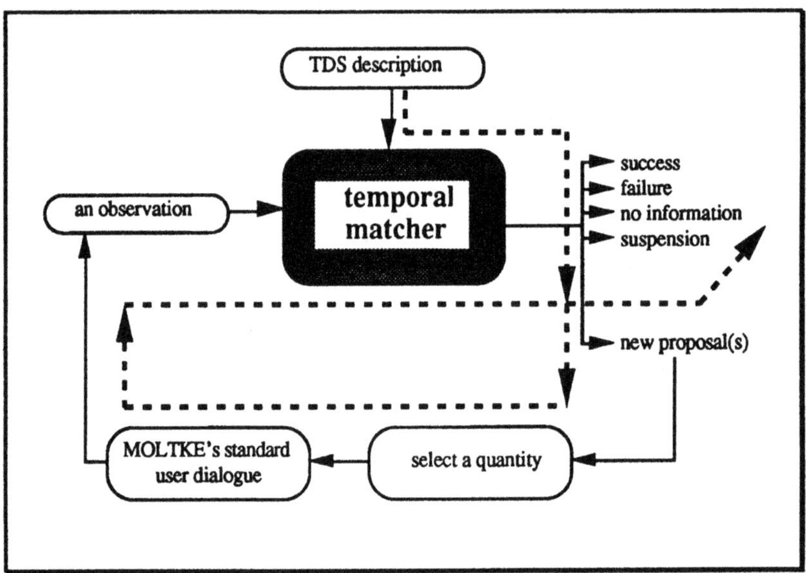

Fig. 60 – The temporal matcher as a black box. The dotted line corresponds to a typical path taken during matching of a TDS

The temporal matcher can be viewed as a black box (fig. 60): once it is initialized with a particular TDS it repeatedly maps an observation to either a suggestion of the next observation or a token that signals one of the three exits from the matching process (i.e. determine the value of the TDS `IOCountPulse`). The matcher is directly derived from fig. 38 by passing through the loop only once. Suggestions are then passed to the same user dialogue mechanism used also for time-independent symptoms which hands back the observed value to the temporal matcher. Thus, after the suggestion of the very first measurement, observing a TDS proceeds as follows: If (due to partial information in the TDS description) there are several equivalent next measurements, the user may choose the quantity he wants to observe. Next he is prompted for the observed value which is matched against the expected value. Depending on the result there are five possibilities four of which have been discussed before: success (correct TDS has occurred), failure (unexpected value), no information (measurement sequence too sparse) or a new suggestion is made. Trying to adhere as closely as possible to the user interface for time-independent symptoms we wanted to preserve the suspension feature described in section 4.1.1. This gives rise to the fifth alternative: each time the user is prompted for a measurement (i.e. several times during the observation of a TDS) he can suspend pursuing the current TDS. During suspension the state slot of the corresponding TDS (26) is changed from `unknown` to `in_progress`. The value of the TDS is not changed (and remains `unknown`). The user can now switch to other contexts and evaluate other rules (including the observation of other TDSs). Afterwards he can return to the original TDS (22) and resume the test at the same stage where it was suspended.

When the matching process terminates, the state (26) and value (27) slots of TDS `IOCountPulse` are filled according to the translation table in fig.61. Notice that with the chosen semantics it is never possible to actually falsify a TDS in MOLTKE.

result of temporal matcher	slot value
success	`true`
failure	`unknown`
no information	`unknown`

Fig. 61 – Relation between result of the temporal matcher and the value slot of the TDS

4.2 Model-Based Diagnosis: GDE / SHERLOCK

Model-based diagnosis[40] is a more recent development than associative diagnosis and has been motivated by some of the inherent limitations of the latter:

- *Availability of experiential knowledge*: The very idea of associative diagnosis presumes that the association between symptoms and diagnoses are known

---

[40] The term "model-based diagnosis" is somewhat misleading. Recently models have also been used, albeit off-line, in the construction of associative diagnostic systems (cf. e.g. [Moissiadis90]). We use "model-based" here in the stricter sense of "simulation-based".

beforehand. Traditionally these associations are derived from the experiential knowledge of maintenance technicians; hence there is an inevitable time lag between the fielding of a new machine and the construction of an associative diagnostic system for it.[41] In view of the decreasing life expectancy of new machine generations this is very problematic, because the machine is likely to become obsolete soon after (or even before!) the diagnostic system is available.

- *Coverage*: For associative diagnostic systems it is very difficult – if not impossible – to estimate the degree of coverage of the faults potentially occurring in the machine. In complex machines new kinds of failures are discovered throughout their life-times and while associative systems can cover the most common faults their performance degrades unpredictably in the presence of unanticipated faults.

- *Multiple faults*: The coverage problem arises because in any sufficiently complex machine not all faults are known. But even when in principle this information is available it may be impractical to construct a complete diagnostic system because of the immense quantity of information. One cause for a combinatorial explosion is the naive treatment of multiple faults in associative systems: since the system has no means of predicting the effects of several interacting faults, every combination of individual faults must be treated as a distinct diagnosis and associations must be collected for each. In the worst case the number of diagnoses thus grows exponentially with the number of individual faults; therefore the majority of associative systems work under the assumption that at most one component is faulted[42].

- *Explainability*: The acceptance of expert systems relies heavily on the users' willingness to believe in their results; a major prerequisite is the system's ability to explain its results in terms of the domain theory. By contrast an associative system can only justify its results in terms of its (domain-independent) inference mechanism; the justifications are grounded in the authority of the technician who is responsible for the domain-specific associations themselves. This may cause

---

[41] Of course, this argument is debatable: only in rare cases machines are totally novel constructions, and in all other cases it may be possible to use experiential knowledge about similar machines at least as a starting point.

[42] Most of these systems have a limited capability of finding multiple faults one at a time by iterating their single-fault strategy under the assumption that the faults interact only weakly.

problems, because the associations may depend on the subjective perspective of the technician.

In its purest form model-based diagnosis is intended to solve all four problems by

- making use only of information about the "correct" behavior of the machine,
- basing its predictions about fault interaction upon simulation in a behavioral model of the machine,
- constructing the behavioral model in terms of universally accepted physical/technical laws (first principles) which can be objectively validated.

In the following section we will briefly describe the GDE system which embodies these principles. When GDE or any other purely model-based diagnostic system is applied to real-world problems it turns out that this approach has its limitations, too. In effect, GDE and other early systems replace the working assumptions of associative diagnosis by new ones; currently much effort is directed at overcoming these novel problems. We emphasize that none of these recent developments directly contradicts the key ideas of model-based diagnosis, or even of GDE. One has to distinguish between model-based diagnosis/GDE as a framework and as a concrete implementation. By nature implementations instantiate a framework in a certain way; we argue that the problems encountered can be at least partially solved, if new implementations exploit the flexibility of the framework. In section 4.2.2 we review one such modification of GDE, SHERLOCK, which is a necessary intermediate step on the way to our integration of TDSs into model-based diagnosis. We then discuss a possible scenario for embedding temporal matching into SHERLOCK and its benefits.

4.2.1 GDE

A detailed description of the General Diagnostic Engine (GDE) developed by de Kleer and Williams can be found in [de Kleer/Williams86] and [de Kleer/Williams87]. In this section we summarize the principal ideas.

GDE operates in two phases (fig. 62):

- In phase 1 (*candidate generation*) the initial set of observations is compared to the behavior the model of the fault-free mechanism. The set of discrepancies is then used to generate the set of all possible diagnoses (candidates) accounting for the discrepancies.

- In phase 2 (*candidate discrimination*) further measurements are proposed to reduce the candidate set until only one is left. The results of the measurements add monotonically to the predictions made by the model and those candidates which fail to explain the additional predictions are eliminated.

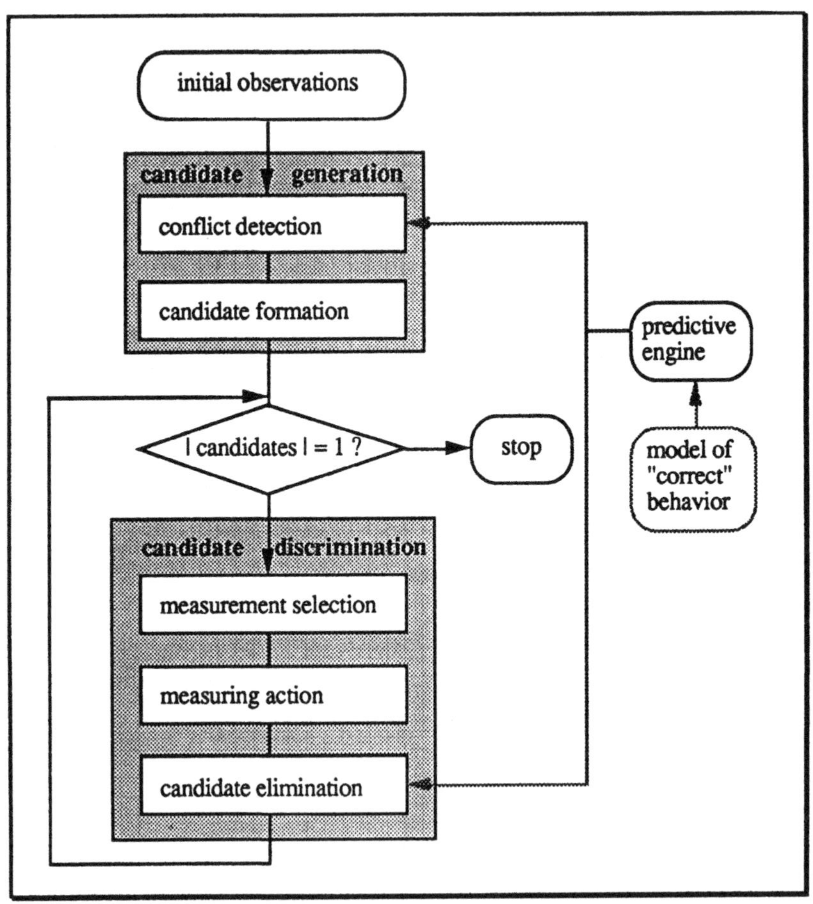

Fig. 62 – The flow diagram for GDE

The refinement of both phases involves a substantial amount of technical effort to control the combinatorial explosion (e.g. the use of an ATMS for recording dependency information during simulation) most of which is irrelevant for our discussion. We will therefore describe the GDE procedure on the conceptual level with the help of a (very familiar) example taken from but being much older than [de Kleer/Williams86] and mention implementation issues only where it is necessary as the basis for chapter 5.

Consider the famous combinatorial circuit in fig. 63 consisting of three multipliers M1, M2, M3 and two adders A1, A2.

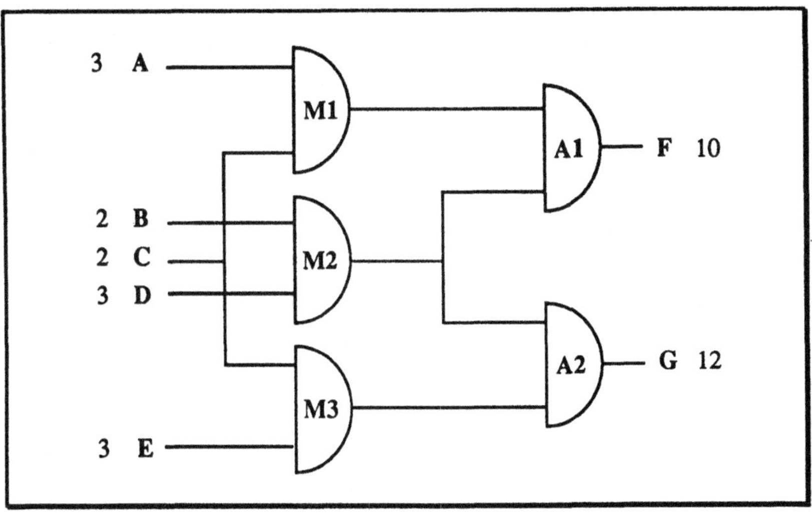

Fig. 63 – The "Tweety" example in model-based diagnosis

Suppose that given the inputs for A - E shown in fig. 63 we have observed the circuit to produce F = 10 and G = 12. Our first task is to detect the discrepancies between this observation and the model of correct behavior for the circuit. The model is simply the set of the five arithmetic equations describing the functionality of the five components. Starting from the observations we can use these equations to predict the values at other points in the circuit. For example, using the equations for M1, M2 and A1 we can calculate that F should be 12 for the given values of A - D, and not 10. Any discrepancy between a prediction and a measurement or between two predictions is called a *symptom*[43]; the set of assumptions leading to a symptom (in our case the correct behavior of A1, M1 and M2) is called a *conflict*. Intuitively, a conflict is a set of assumptions at least one of which must be false. Since any superset of a conflict is also a conflict we can characterize the set of all conflicts by computing the set of all *(subset-) minimal conflicts*. This set can be used to canonically generate the set of diagnoses that account for the observed discrepancies. Diagnoses are sets of assumptions all of which must be false. They can be approximated by the set of *candidates*. Every candidate must

---

[43] Notice that the meaning of "symptom" in GDE is different from its usage in MOLTKE!

explain all of the symptoms; hence a candidate must have a non-empty intersection with each minimal conflict. Diagnoses then correspond to the *minimal candidates*, i.e. those candidates which contain *exactly* those assumptions needed to explain the symptoms. In our example there are two minimal conflicts {A1, M2, M1} and {A1, A2, M1, M3} producing as minimal candidates the sets {A1}, {M1}, {A2, M2} and {M2, M3}.

Since there is more than one minimal candidate, we have to propose further measurements to home in on the actual diagnosis. Candidate discrimination can be further decomposed (fig. 62) into selecting the next observation and interpreting its result. The first design decision concerns the set of potential measurements from which we choose. GDE's policy in this respect seems somewhat arbitrary and probably reflects the properties of circuit diagnosis which is GDE's standard domain. First, GDE assumes that all components are accessible and that each of the component terminals can be measured. Secondly, all measurements are equally (in-)expensive so that test costs are not a criterion. Thirdly, all measurements throughout the diagnostic session take place within the same state of the machine – the state in which the initial observations were made. In particular the set of inputs is never varied.[44] The last assumption is largely a matter of efficiency since conflict detection and candidate generation can be optimally supported by the underlying ATMS when the new observation can be added incrementally to the same machine state (in GDE terminology: environment). However, the first and third assumptions together imply that GDE is helpless when in practice a measurement between two suspected components is impossible or (contrary to the second assumption) too expensive. The problem does not even disappear with the introduction of models of the faulty behavior in SHERLOCK (in section 4.2.2) and is the prime motivation for the extension we propose in section 4.2.3.

In the basic GDE all component terminals are potential places for measurements and the planning horizon is confined to one measurement. The selection between potential observations is based entirely on the expected information gain. It is assumed that components fail independently of each other and that all components c have an *a priori*

---

[44] This approach to diagnosis which can be characterized as gradually exploring one device state is commonly called *probing*.

probability $p_a(c)$ of being broken[45]. Then the initial probability of a candidate K is given by

$$p_a(K) := \prod_{c \in K} p_a(c) \cdot \prod_{c \notin K} (1 - p_a(c))$$

The candidate probabilities change each time a new observation is added: those candidates which are incompatible with the observation are eliminated (i.e. their probability becomes zero) and their probability mass is redistributed between the remaining candidates. We will take a closer look at the update step in section 4.2.3.1.

Based on the candidate probabilities the amount of information at a particular stage of the diagnostic process can be measured by Shannon's *entropy* function

$$H = - \sum_{i} p(K_i) \log p(K_i)$$

The utility of a potential measurement can then be estimated by computing the expected decrease in entropy. Suppose that a potential next action is to measure quantity q and that the set of possible outcomes is $\{v_1, \dots, v_m\}$[46]. One can compute the a posteriori probabilities $p(q = v_k)$ for each outcome v_k and from these the updated candidate probabilities and the resulting new entropies $H(q = v_k)$. Thus the expected entropy $H_e(q)$ after measuring q is given by

$$H_e(q) = \sum_{k=1}^{m} p(q = v_k) \, H(q = v_k)$$

It is therefore best to measure a quantity q_{opt} which maximizes

$$\Delta H_e(q) := H - H_e(q).$$

---

45 Under the presuppositions of equiprobable faults and very small fault probability this approach can also be used when the a priori probabilities are not known (cf. [de Kleer89]).

46 Recall that GDE talks about *static* systems, i.e. the predicted "behavior" consists of only one value per quantity.

4.2.2 SHERLOCK

In contrast to associative diagnostic systems whose knowledge bases contain primarily information about abnormal behavior of a device GDE's only source of information is a model of the correct behavior (*physiological model*). This is motivated by the claim that the physiological model can be extracted "from the blueprints" in contrast to information about malfunctions which is available only after experience and that therefore a diagnostic system based on a physiological model will be available earlier in the life-cycle of a machine. But there are also severe disadvantages. First, a diagnosis pinpoints only the faulted components without specifying *in which way* they fail. This may be insufficient to select appropriate repair actions. Second, the initial candidate failure probabilities include all possible misbehaviors without further differentiation. In domains where the independent failure assumption is violated it is not possible to assign individual probabilities to combinations of *particular* misbehaviors. Third, the prediction phase is weak because adding a component to a candidate means retracting all its behavioral axioms from the model (rather than replacing them with an alternative set of axioms). As a consequence a component that is believed to be faulted can behave in absolutely any way, which leads to extremely awkward results. Consider e.g. the circuit in fig. 64 in which two light bulbs L1, L2 are connected in parallel to a battery B. Suppose we observe that L2 is lit and L1 is not. Apart from the correct candidate {L1} GDE produces spurious candidates such as {B, L2} which can be interpreted as "B is faulted and produces no current and L2 is faulted and is lit without current". Naturally, a more realistic model of a light bulb – i.e. one that specifies in which ways light bulbs can be broken – would exclude this candidate.

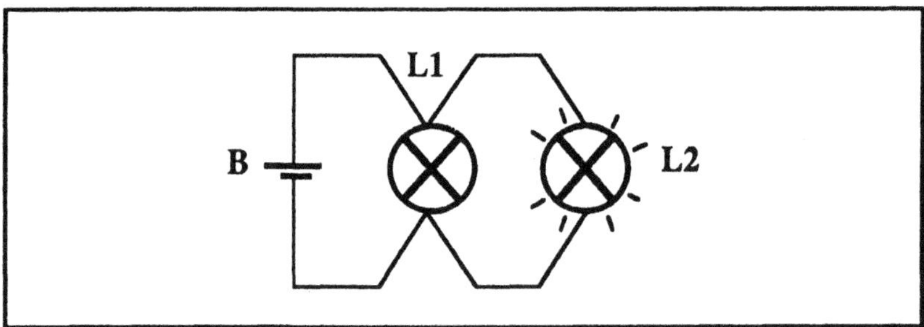

Fig. 64 – The battery-and-light-bulb example from [Struss/Dressler89] (slightly modified)

The situation gets worse when one tries to apply the GDE framework to dynamic systems, as Hamscher and Davis pointed out already in [Hamscher/Davis84].

The logical consequence is SHERLOCK [de Kleer/Williams89], GDE's successor, which attempts to reconcile the principles of model-based diagnosis with the use of fault models[47]. In SHERLOCK a component c may be not only in one of two states (ok or faulted), but in one of an arbitrarily large set of behavioral modes {OK(c), F_1(c), … , F_n(c), U(c)} where F_1, … , F_n correspond to particular known malfunctions and all potential, but unknown faults are lumped together in U. Each state of a component has its own a priori probability with p(U(c)) typically being very small. Furthermore, models exist not only for OK(c) but also for each F_j(c). U(c) takes the place of the former "faulted-and-not-knowing-anything-about-how"; it has no model. In SHERLOCK's framework diagnosis is viewed as the task of determining for each component the behavioral mode it is in. Most of GDE's terminology is carried over, only the meaning of candidates is reversed: a candidate is now a set of assumptions about behavioral modes *all* of which are believed to be *true*. In the light bulb example the physically correct explanation for the observations would thus be the candidate {OK(B), BURNOUT(L1), OK(L2)}. The majority of changes from GDE to SHERLOCK are technical and do not concern us here. Conceptually, selecting measurements and updating candidate possibilities is carried out exactly as in GDE with the single important difference that more of the candidate models – and virtually all of the high-probability ones – are specified completely so that more quantity values can be predicted and more candidates tend to be ruled out with each measurement.

4.2.3 Integrating TDSs into SHERLOCK

To date GDE and SHERLOCK have been used almost exclusively in domains which can be modeled statically[48]. In fact, in their original form both systems are unable to cope with several inherent properties of fault diagnosis in dynamic systems:

1) The set of potential "places" for measurements is exponentially larger than in the static case, because measurements may involve the concurrent observation of *several* quantities. Therefore in the general case it is not practical to explicitly compute expected decreases in entropy for all of them.

---

[47] models of incorrect behavior

[48] A notable exception is the system described in [Decker89] about which we will say more in section 4.2.3.3.

2) The measurement outcomes are correspondingly more complex; instead of just one value we get a measurement sequence.

3) The predicted "values" for the measured quantities are now behaviors over extended periods of time instead of atomic, time-independent values.

4) The set of single-time-point, single-quantity measurements yielding non-zero information gain may be exhausted before only one candidate is left, because some candidates may be distinguishable solely on the basis of TDSs.

5) The concept of external actions (as in dynamic experiments) is absent from GDE / SHERLOCK's measurements.

We will now describe an extended framework based on SHERLOCK which is intended to solve problems 2) – 4). In the next section we will explain why on the contrary a universal and practical solution to problem 1) – especially in combination with 5) – is not to be expected in the near future. Furthermore we argue that this latter problem possesses more aspects of a planning situation than of diagnosis; it is therefore for the greater part outside the scope of this thesis.

In our discussion we will refer to SHERLOCK as described by de Kleer and Williams as *static SHERLOCK* and to our extended framework as *dynamic SHERLOCK*. We divide the presentation into three parts:

- In section 4.2.3.1 we describe on the conceptual level how temporal matching can be incorporated into SHERLOCK as a generalized form of measurement and how this affects conflict detection and candidate probability updating.
- In section 4.2.3.2 we augment static SHERLOCK's algorithm for candidate discrimination so that the generalized measurements of section 4.2.3.1 are exploited.
- We illustrate the algorithm in section 4.2.3.3 where we give an example of its operation, demonstrating how a problem from the literature can be attacked in a systematically more satisfying way than before.
- Finally, we evaluate our approach in section 4.2.3.4.

4.2.3.1 Generalized Measurements – Theory

We start by describing static SHERLOCK's approach to candidate discrimination and propose a generalization of the computation of candidate probabilities and expected entropies that encompasses temporal matching and static SHERLOCK's individual measurements as special cases. We will first give an informal comparison of the two approaches.

Common to static and dynamic SHERLOCK is the fact that a place for a measurement is selected, each candidate predicts one or more possible outcomes and the candidate probabilities are updated depending on the actual result of the measurement. The differences concern the type of model, the notion of "a place for a measurement" and the form of predicted and actual measurement results; they are summarized in fig. 65.

concept	static SHERLOCK	dynamic SHERLOCK
model type	static	dynamic
"places" for measurements	a quantity q to be observed (the system state never changes and is therefore irrelevant)	a set Q of quantities to be observed over a period of time starting in a particular system state[49]
predictions	for each candidate a value from dom(q)	for each candidate[50] a set of behaviors (TDS descriptions) for the quantities in Q
outcomes	a value from dom(q)	a measurement sequence in response to a measurement plan suggested by running the temporal matching algorithm in parallel for all predicted behaviors.

Fig. 65 – Principal differences between static and dynamic SHERLOCK

---

[49] This state may be different from measurement to measurement. Thus by varying the "inputs" in order to bring about characteristic reactions in the "outputs" it is possible to introduce an element of *testing* into SHERLOCK.

[50] We will come to the question of candidates containing components with unknown fault modes (U-modes) in a moment.

To avoid confusion we reserve the term "measurement" for the individual constituents of a measurement sequence (as in chapter 3) and for static SHERLOCK's observations and use the term "generalized measurement" when we speak of groups of measurements forming one step in dynamic SHERLOCK's procedure.

Already the informal comparison in fig. 65 allows some interesting observations. Most important, at any stage of the diagnostic process in dynamic SHERLOCK the set of potential "places for measurements" is much greater than in the static case. Of course, the vast majority of these generalized measurements does not contribute new information about the candidates, but there is no straightforward way of singling out those that do. Unfortunately, this means that it is impossible to use the same selection strategy as in static SHERLOCK where all potential measurements are explicitly enumerated and the one with maximum expected decrease in entropy is selected. In dynamic SHERLOCK the sheer number of potential measurements precludes this approach.

An ambitious solution to this problem reverses the whole process, starting with the desired effect (i.e. a selection of the candidates to be eliminated next) and constructing a generalized measurement specifically for this effect. We will discuss this approach in a little more detail in section 5.3 where we will point out the primary obstacles that at the time of writing make it appear interesting, but still prohibitively expensive. These obstacles are mainly due to the deficiencies in the state of the art of the supporting techniques (e.g. planning, qualitative simulation).

Fortunately though, there are also cases[51] in which the set of useful, potential generalized measurements is not only small, but even known. Such information may be available from the body of general background knowledge in a subfield of engineering, even if no specific experience with the particular device has yet been gathered. If this is true in a given application, we are basically in the same situation as in static SHERLOCK. Our problem is then reduced to the following tasks:

- representing generalized measurements,
- representing the predicted measurement outcomes for each candidate,
- computing the a posteriori candidate probabilities for each outcome,

---

[51] One of these cases will serve as our example in section 4.2.3.3.

- specifying how the individual measurements in a generalized measurement are to be performed.

We will now describe our solution formally, the only exceptions being the concept of device state and the prediction mechanism itself. Just as in the state-oriented approach to behavior representation described in section 3.2.1 we appeal to the intuition that the *state* of a device can be characterized by the values of certain quantities (sometimes also called *state variables*).

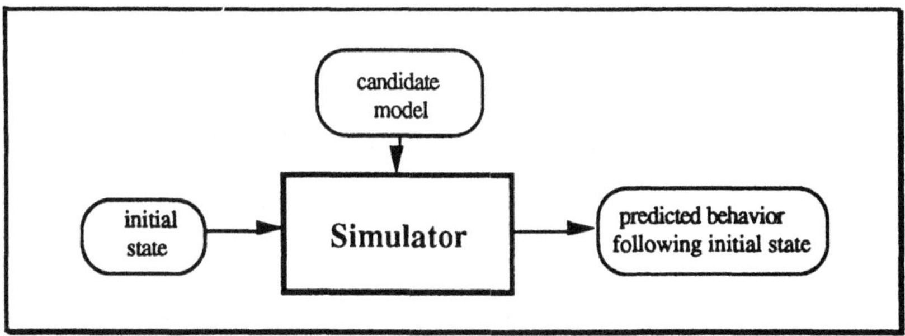

Fig. 66 – Input / output requirements of the simulator

Our second assumption concerns the prediction mechanism: we require that it is possible to predict the behavior of the device under the hypotheses of a candidate using some kind of simulation in the candidate model. We do not put any restrictions on the predictive engine as such; all we demand is the following (fairly general) input/output relation (fig. 66): given the candidate model and an initial state of the device, the simulator produces a description of the device behavior for some period following the initial state. The length of the period depends on the time it takes for the fault to manifest itself in the behavior of observable quantities. There are several possible implementations of this criterion. In a demand-driven version the diagnostic system would examine the simulator output regularly and would stop the simulation as soon as the results are sufficient for discrimination. In fact, the algorithm for preparing a generalized measurement that we develop in the next section could be turned into such a decision procedure.

The output format varies from simulator to simulator, but a translation between formats is usually possible[52]. Without loss of generality we assume that the simulator output

---

[52] If formats differ in their level of abstraction, the translation is at least possible in the direction from the more detailed to the more abstract description.

can be translated into the TDS description format. This assumption is generally unproblematic for history-based (e.g. HIQUAL [Voß86]) and state-based (e.g. QSIM [Kuipers86]) outputs of qualitative simulators; for quantitative simulators it may be necessary to abstract from real-valued, continuous quantities to obtain a behavior description in terms of histories in which the episodes have non-zero extent.

Due to the inevitable information loss in the abstraction process qualitative simulators do not always predict only the physically real behavior of the device, but in addition a number of spurious behaviors. Some of these alternative behaviors can be represented in a single TDS description using disjunctive interval relations. However, as we have demonstrated in section 3.2.4, due to the limitations of Allen relations this is not always possible. Allowing for ambiguous predictions in the general case we define our abstract view of the simulator in the form of a function, which for a candidate, a set of quantities and a fixed initial state computes the device behavior in the form of a *set* of TDS descriptions.

DEFINITION 23: The *behaviors for a candidate K, a set of quantities Q, starting in the initial state IS* is given as the set

$$Beh(K,IS,Q) = \{ S_1, \ldots, S_k \}$$

where the S_i are TDS descriptions of the form $\langle Q, H_i, C_i \rangle$.

A problem arises when components with U–modes are present in a candidate. While it is difficult to predict a behavior for partially specified *static* models, it is nearly impossible in the *dynamic* case. We therefore assign to such candidates the set of all behaviors Ω and declare that all measurement outcomes are compatible with Ω.

In selecting a generalized measurement in dynamic SHERLOCK we start from an initial device state, a set of quantities to be observed and the sets of behaviors predicted for each candidate. To synthesize a generalized measurement from this information we must first plan a sequence of observations in such a way that we can detect which of the predicted behaviors actually occurs. Next we have to estimate the utility of the generalized measurement by computing the sets of measurement sequences which are compatible with the behaviors predicted by each candidate and from these the resulting a posteriori candidate probabilities for each outcome.

Formally:

DEFINITION 24: A *measurement plan* MP(Q) for a set of quantities Q is a set $\{\langle q_i, t_i \rangle\}_{i=1,\ldots,n}$ where $t_1 < t_2 < \ldots < t_n$, $q_i \in Q$.

DEFINITION 25: A *generalized measurement* is a pair ⟨IS,MP(Q)⟩ where

- IS is a system state and
- MP(Q) is a measurement plan.

Outcomes of generalized measurements are the same measurement sequences that we introduced in section 3.4. Just as measurements in static SHERLOCK can produce a value from the domain of the observed quantity, a measurement plan induces a set of possible sequences of results.

DEFINITION 26: The set of *possible measurement sequences* for a measurement plan $MP(Q) = \{\langle q_i, t_i\rangle\}_{i=1,\dots,n}$ is the set $AllSeq(MP(Q)) := \{ \{\langle q_i, t_i, v_i\rangle\}_{i=1,\dots,n} \mid v_i \in dom(q_i) \}$.

Notice that in general

$$| AllSeq(MP(Q)) | = \prod_{\langle q_i,t_i\rangle \in MP(Q)} | dom(q_i) |,$$

although not all sequences may be physically possible.

Now we can state which of the possible measurement sequences are compatible with the behavior predicted by a particular candidate. Again the definition is a straight-forward generalization: instead of the simple comparison of atomic values in static SHERLOCK we have to check whether the observed measurement sequence is compatible with one of the possible behaviors of the candidate.

DEFINITION 27: The set of measurement sequences $M \in AllSeq(MP(Q))$ which are *compatible* with candidate K and initial state IS is defined as
$Seq(K,IS,MP(Q)) := \{ M \in AllSeq(MP(Q)) \mid S \in Beh(K,IS,Q),$
M compatible[53] with an instance of S $\}$.

In particular, if $Beh(K,IS,Q) = \Omega$, then by definition Seq(K,IS,MP(Q) is equal to AllSeq(MP(Q)) for all Q and MP(Q).

Under the assumption that a particular measurement sequence from AllSeq(MP(Q)) has been observed we can compute the new probability for each candidate K. Let

---

[53] in the sense of definition 16 in section 3.4

"$\langle IS,MP(Q)\rangle \rightarrow M_k$" denote the event in which the measurement sequence M_k is the outcome of the generalized measurement $\langle IS,MP(Q)\rangle$. Using Bayes' rule we obtain the conditional probability

$$(*) \qquad p(K \mid \langle IS,MP(Q)\rangle \rightarrow M_k) = \frac{p(\langle IS,MP(Q)\rangle \rightarrow M_k \mid K)\ p(K)}{p(\langle IS,MP(Q)\rangle \rightarrow M_k)}$$

Here p(K) is K's probability after the last discrimination step (resp. its a priori probability), so that we only need to determine the other two factors. Like de Kleer and Williams we assume (sometimes without justification) that if more than one measurement sequence is compatible with a candidate, all sequences occur with equal probability. We thus obtain:

$$p(\langle IS,MP(Q)\rangle \rightarrow M_k \mid K) = \begin{cases} 0, & \text{if } M_k \notin Seq(K,IS,MP(Q)) \\ \frac{1}{|Seq(K,IS,MP(Q))|} & \text{otherwise} \end{cases}$$

The denominator of (*) is a scaling factor that does not depend on K. Following the same argument as above we get

$$p(\langle IS,MP(Q)\rangle \rightarrow M_k) = \sum_{K:\ M_k \in Seq(K,IS,MP(Q))} \frac{p(K)}{|\ Seq(K,IS,MP(Q))\ |}$$

We can finally substitute these formulae into (*) obtaining

$$(**)\ p(K \mid \langle IS,MP(Q)\rangle \rightarrow M_k) = \begin{cases} 0, & \text{if } M_k \notin Seq(K,IS,MP(Q)) \\ \frac{p(K)}{p(\langle IS,MP(Q)\rangle \rightarrow M_k)\cdot|Seq(K,IS,MP(Q))|} & \text{otherwise} \end{cases}$$

The argumentation leading to (**) can be backed up by taking a look at the updating step in static SHERLOCK and checking whether it is indeed a special case of this formula. In this case the initial state IS is irrelevant, because the system state does not change during the diagnostic session. MP(Q) consists of exactly one observation of the quantity x_j (the observation time does not matter) and M_k contains the measured value $v_{jk} \in dom(x_j)$. Due to static SHERLOCK's constraint propagation strategy candidates always predict either exactly one value or none at all (in the second case they are called ***uncommitted***), i.e. either $|Seq(K,IS,MP(Q))| = 1$ or $|Seq(K,IS,MP(Q))| = |AllSeq(MP(Q))|$. Under this presupposition about the predictive engine the formula given in [de Kleer/Williams87] corresponds precisely to (**). Incidentally, if we generalize de Kleer and Williams' formula to non-singleton predictions, we obtain (**), too.

4.2.3.2 An Augmented Algorithm for Candidate Discrimination

How can we use the theoretical framework proposed in the previous section to redesign SHERLOCK's algorithm for candidate discrimination? Fig. 67 is a blow-up of the relevant portion of fig. 62, annotated with the information flow between subsequent stages of candidate discrimination.

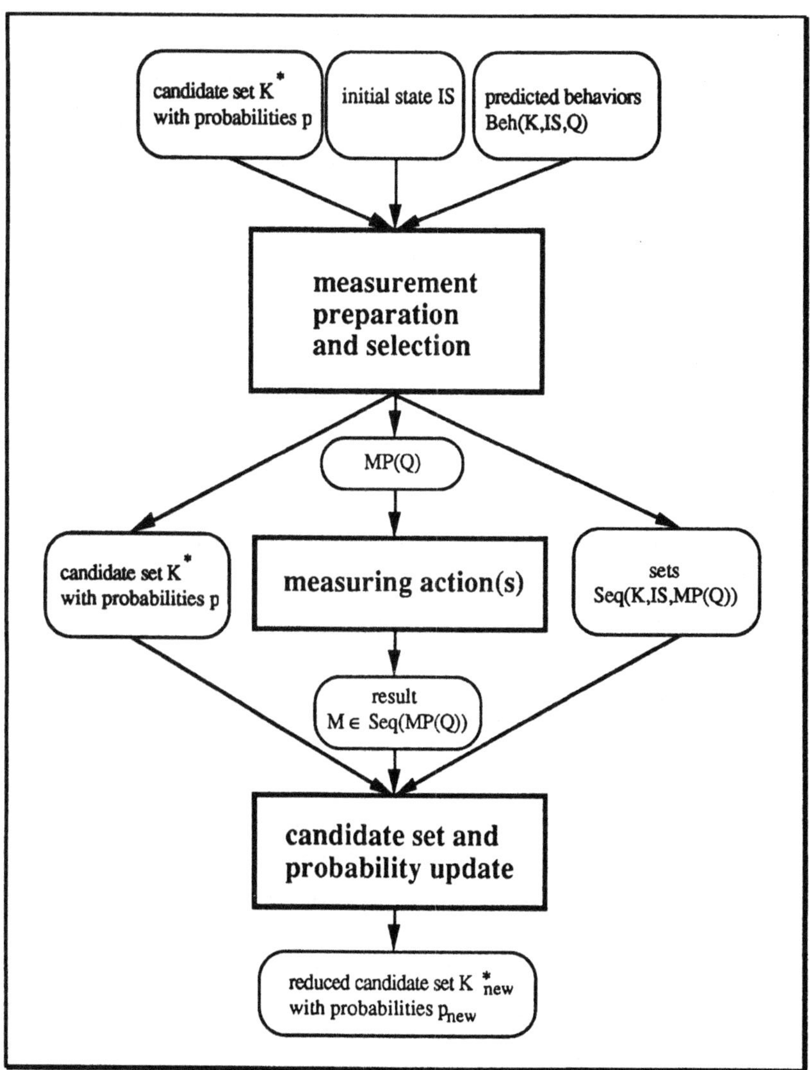

Fig. 67 – Detailed view of candidate discrimination in dynamic SHERLOCK

As we can see, candidate discrimination proceeds in three stages:

1) In the *selection phase* we compute the expected information gain (entropy decrease) for each potential measurement and select an optimal measurement. For static measurements this is done exactly as in static SHERLOCK; for measurements based on predictions of dynamic behaviors we must determine a measurement plan MP(Q) and the sets Seq(K, IS, MP(Q)) for $K \in K^*$, before we can apply equation (**) from above.

2) Next the *measurement actions* specified by MP(Q) are carried out yielding a result $M \in$ AllSeq(MP(Q)). Again, in the static case MP(Q) specifies only one measurement and M contains only one (atomic) result.

3) Finally, in the *update step* all candidate probabilities are recomputed according to the measurement result M. Candidates which receive the probability 0 are removed from K^* leading to a new candidate set K^*_{new}.

We will first refine the selection phase which is the most complicated of the three. Since static measurements are treated as in static SHERLOCK, we only consider generalized measurements based on dynamic behavior prediction. Conceptually, we proceed in three steps for each general measurement:

1.1) At the beginning, only the initial state IS and the behaviors Beh(K, IS, Q), $K \in K^*$, for some set Q of quantities are known. From this information we construct a measurement plan MP(Q) which together with IS will form the generalized measurement ⟨IS, MP(Q)⟩.

1.2) For each candidate $K \in K^*$ we determine the set Seq(K, IS, MP(Q)), i.e. the set of those outcomes of ⟨IS, MP(Q)⟩ which are compatible with K.

1.3) Using the sets Seq(K, IS, MP(Q)) we can compute the a posteriori candidate probabilities for each outcome $M \in$ AllSeq(MP(Q)) according to equation (**) and from this the expected a posteriori entropy H_e(⟨IS, MP(Q)⟩).

We can then select the measurement with the minimal expected entropy.

Several aspects have an influence on the design of an algorithm for 1.1) - 1.3). First, recall that for candidates with U-modes the simulator predicts Ω, the set of *all* possible behaviors. These candidates do not contribute any useful information to the construction of MP(Q), because any measurement sequence obtained in response to any MP(Q) is compatible with them. Secondly, the measurements specified by MP(Q) must be sufficiently numerous and densely spaced. Definition 16 implies that short and/or sparse measurement sequences are compatible with almost any TDS instance, because most or all measurements fall outside the basis interval of the instance. Measurement sequences which can be expected to differentiate between potential TDS instances (by being *not* compatible with most of them) must at least contain a subsequence with the granularity properties in the precondition of theorem 8 (section 3.5). Consequently, MP(Q) has to be constructed in such a way that the measurement sequences in AllSeq(MP(Q)) have these properties. Theorem 8 contains one possible solution: if MP(Q) is planned according to the properties stated there, the resulting measurement sequence must either match a given TDS or not be compatible at all. In the case of a single TDS description we could take the ordinary temporal matching algorithm from section 3.6, let it plan MP(Q) and determine those measurement sequences that lead to a successful match. In the case at hand, where we want to differentiate between *several* competing TDSs, we can use a parallel version of the same set-up. We run the temporal matching algorithm concurrently on all predicted TDSs, take the union of the measurement suggestions produced for each TDS individually and record which observed values lead to a match. Notice that all of this happens in a kind of thought experiment in the selection phase, *before* any measurements are actually carried out in the measuring action phase. The procedure that we have just sketched is formalized in the following algorithm PREPARE_AND_EVAL_MSMT.

```
Algorithm PREPARE_AND_EVAL_MSMT

Input:        a set K* of candidates,
              the current candidate probabilities p(K), K ∈ K*,
              an initial state IS,
              a set of quantities Q,
              the sets Beh(K,IS,Q), K ∈ K*.
Output:       A measurement plan MP(Q) for the generalized measurement
              ⟨IS,MP(Q)⟩, derived from the sets Beh(K,IS,Q),
              the sets Seq(K,IS,MP(Q)), K ∈ K*,
              the expected entropy He(⟨IS,MP(Q)⟩).
Temporary data structures:
              S*: a set of TDSs,
              T, NewT, Tsuccess, Tfinished:
                     sets of triples ⟨S,G,M⟩ where
                     S is a TDS, G is a record containing the data
                     structures for the ordinary temporal matching
                     algorithm (reality time net, open, sleeping,closed)
                     and M is a measurement sequence (such a triple is
                     called an instantiation record),
              Q*: a set of quantities,
              V(q): for each q ∈ Q a set of values.
```

begin

(1) $S^* \leftarrow \bigcup_{\substack{K \in K^* \\ Beh(K,IS,Q) \neq \Omega}} Beh(K,IS,Q);$

(2) $MP(Q) \leftarrow \emptyset;$

(3) $T_{finished} \leftarrow \emptyset;$

(4) run the temporal matching algorithm MAIN concurrently
on all $S \in S^*$ as follows (steps (5) - (25)):

(5) for each $S \in S^*$ initialize the data structures $G(S)$;

(6) $T \leftarrow$ the set of all triples $\langle S, G(S), \emptyset \rangle$;

(7) Run the algorithm SUGGEST for each instantiation $\langle S,G,M \rangle \in T$.
Let Q^* be the union of all suggested quantities.
For each $q \in Q^*$ let $V(q)$ be the set of all values expected to
be measured next for q. If a quantity q is not suggested by
all instantiations, then $V(q) \leftarrow dom(q)$.

(8) $T_{success} \leftarrow$ the set of $\langle S,G,M \rangle \in T$ for which no new measurements
have been suggested and where the reality time net is a
specialization of the TDS time net ;

(9) $T_{finished} \leftarrow T_{finished} \cup T_{success};$

(10) $T \leftarrow T \setminus T_{success};$

(11) if $T = \emptyset$, then goto (26);

(12) assume $Q^* = \{q_1,\dots,q_s\}$, $s \geq 1$, (the numbering is arbitrary);

(13) append $\{\langle q_1,t_1 \rangle,\dots,\langle q_s,t_s \rangle\}$ to $MP(Q)$, where t_i are arbitrary
time points in ascending order which are all greater than
the time points picked on previous turns and which satisfy
the granularity properties required by theorem 8;

(14) $NewT \leftarrow \emptyset;$

(15) for each sequence $\langle v_1,\dots,v_s \rangle \in V(q_1) \times \dots \times V(q_s)$:

(16)   for each $\langle S,G,M \rangle \in T$:

(17)   match the measurements $\langle q_1,t_1,v_1 \rangle,\dots,\langle q_s,t_s,v_s \rangle$ against
S and G according to algorithm MATCH;

(18)   if the match succeeds,

(19)   then $NewT \leftarrow NewT \cup \{\langle S,G',M' \rangle\}$ where
G' contains the updated data structures and
$M' = M \cup \{\langle q_1,t_1,v_1 \rangle,\dots,\langle q_s,t_s,v_s \rangle\};$

(20) $T_{finished} \leftarrow \{\langle S,G,M' \rangle \mid \langle S,G,M \rangle \in T_{finished},$
$\langle v_1,\dots,v_s \rangle \in dom(q_1) \times \dots \times dom(q_s),$
$t_1,\dots,t_s$ as above in step (13),
$M' = M \cup \{\langle q_1,t_1,v_1 \rangle,\dots,\langle q_s,t_s,v_s \rangle\}\ \}$

(21) $T \leftarrow NewT;$

(22) goto (7);

(23) for each $K \in K^*$:

(24)   if $Beh(K,IS,Q) \neq \Omega$

(25)   then $Seq(K,IS,MP(Q)) \leftarrow \bigcup_{S \in Beh(K,IS,Q)} \ \bigcup_{\langle S,G,M \rangle \in T_{finished}} M$

(26)   else $Seq(K,IS,MP(Q)) \leftarrow AllSeq(MP(Q));$

(27) compute $H_e(\langle IS,MP(Q) \rangle)$ according to equation (**) using

p(K) and Seq(K,IS,MP(Q)).

end.

Comments:

(1) S^* is the set of *all* TDS descriptions that we want to discriminate between. Candidates which do not predict any particular behavior (Ω) receive special treatment in step (26) and are not considered here.

(3) Not all TDS descriptions in S^* need equally long measurement sequences to be detected. At any stage of the algorithm $T_{finished}$ contains those TDS descriptions (along with additional information) which have already been matched successfully by measurements specified in an initial segment of the current MP(Q).

(5) according to steps (1) - (5) of algorithm MAIN.

(6) Each instantiation record corresponds to one call to the ordinary temporal matching algorithm. The significance of S and G(S) is obvious. The third element of each instantiation record contains a measurement sequence $M \in$ AllSeq(MP(Q)) for the current MP(Q) which is being extended to a measurement sequence that matches S.

(8) In these cases M matches S according to algorithm MAIN.

(10) - (11) From here on we consider only those TDS description which need further measurements. If there are none left, we are done. In the last case all instantiations that have ever been created in steps (6) or (20) have either been moved to $T_{finished}$ in steps (9)-(10) or been eliminated due to a mismatch in steps (17)-(18).

(12) If $T \neq \emptyset$, then $Q^* \neq \emptyset$.

(13) If $\{q_1,\dots,q_s\}$ are the quantities suggested by the individual instantiations of the temporal matching algorithm, we plan to observe them all in arbitrary order.

(15) For every outcome of the new measurements that corresponds to an expected value in any of the instantiations, perform the MATCH step for each instantiation.

(17) - (18) Perform MATCH in turn for each measurement. If after processing measurement $\langle q_i,t_i,v_i\rangle$ the TDS description S has been matched completely, then ignore the rest of the measurements $(\langle q_j,t_j,v_j\rangle,\ i<j\leq s)$.

(19) Notice, first, that for each $\langle S,G,M\rangle \in T$ *several* new instantiations $\langle S,G',M'\rangle$ may be added to NewT, if more than one outcome of the new measurements is compatible with S and G. Secondly, if the match is not successful, *nothing* is added to NewT. This is the second place (after step (10)) where instantiations are effectively removed from T.

(20) The TDS descriptions in $T_{finished}$ have already been matched completely on a previous turn. For them the outcomes of later measurements are insignificant. To get measurement sequences of equal length for all $S \in S^*$ we extend the measurement sequences in $T_{finished}$ with *all* possible outcomes of the new measurements in MP(Q)).

(23)-(26) Determine the sets Seq(K,IS,MP(Q)). If Beh(K,IS,Q) is made up of TDSs in S^*, we retrieve the information from the measurement sequences belonging to successful matches in $T_{finished}$. If the behavior is unspecified (because K contains a component in a U-mode), set Seq(K,IS,MP(Q)) explicitly to the set of all possible outcomes.

Obviously, the most expensive part of PREPARE_AND_EVAL_MSMT is the computation of the sets Seq(K, IS, MP(Q)). Fortunately, the correctness of this computation does not depend on the current candidate probabilities at all and – as long as K^* is reduced monotonically – not on the set K^* as a whole. It is therefore possible to reuse the sets in later discrimination steps without the risk of incorrect conclusions. The only disadvantage concerns efficiency: since in later steps the candidate set and hence the set of predicted behaviors is smaller, fewer measurements may suffice to differentiate between them. Nevertheless MP(Q) and the sets Seq(K, IS, MP(Q)) are still valid for discrimination, albeit perhaps redundant.

When all expected a posteriori entropies are known, an optimal measurement is selected and carried out. The result is a measurement sequence $M \in$ AllSeq(MP(Q)) which can be used to define

$$K^*_{new} := \{K \in K^* \mid M \in Seq(K,IS,MP(Q))\},$$

the set of candidates with which the measurement outcome is compatible. K^*_{new} becomes the new candidate set at the beginning of the next discrimination step. All that remains to be done is to update the candidate probabilities for $K \in K^*_{new}$ according to equation (**).

4.2.3.3 Example

In this section we demonstrate the operation of dynamic SHERLOCK with the help of an extended example. The device that we consider is a thyristor circuit which is used in various forms in power electronics (including e.g. motor control in CNC machines). The reasons for selecting thyristor circuits are twofold:

- These devices possess the properties which motivate our extension: a predictable behavior over time which is characterized more by the temporal relations between the episodes than by unique episode values, and faults which can be diagnosed exclusively using TDSs.

- Thyristor circuits have been used before in the literature ([Decker89], [Struss90]) as reference examples for model-based diagnosis of dynamic systems. They are therefore a particularly good test case for new approaches to the treatment of TDSs.

We will first provide some condensed background information about thyristor circuits[54]. Then we describe a typical situation during fault diagnosis of a thyristor circuit in which no static measurement can further reduce the candidate set, and show how a generalized measurement solves the problem.

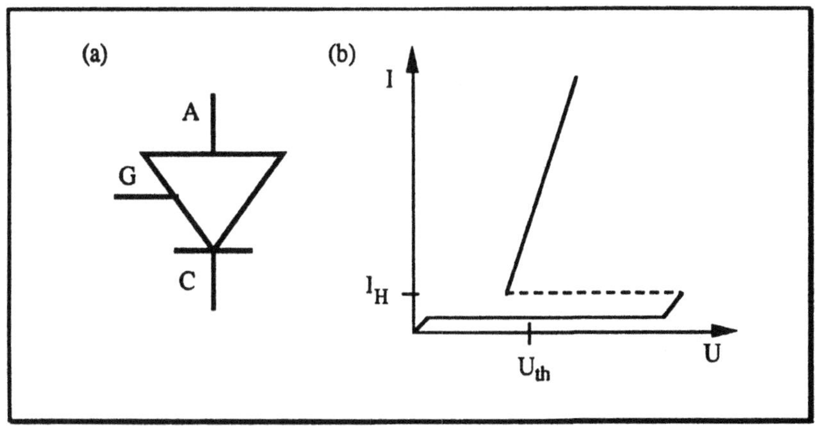

Fig. 68 – The thyristor

A *thyristor* (fig. 68 (a)) is a semi-conductor with anode, A, cathode, C, and gate, G. It can operate in two states, either conducting current in a specified direction with an almost zero resistance, or working as a resistor with an almost infinite resistance, as is indicated by the characteristic curve in fig. 68 (b). The transition from the *blocking* to the *conducting* state is controlled by the gate: If G receives a pulse from a control unit at time t_0 and if the voltage drop across the thyristor is above a threshold, $\Delta U(t_0) > U_{th}$, the thyristor is fired and conducts current. The thyristor reverts to the blocking state, when the current through the thyristor drops below another threshold I_H. Thus, in a first approximation, the thyristor operates as a directed switch. This model suffices to understand the basic functionality of the circuit shown in fig. 69.

In this circuit S is a source that provides three-phase current (voltages A, B, C), T_1, ..., T_6 are thyristors, and M is a motor. The circuit serves two distinct purposes. First, it rectifies alternating current, since current is only directed in one direction and, hence,

---

[54] This description is taken from [Struss90] with kind permission of the author.

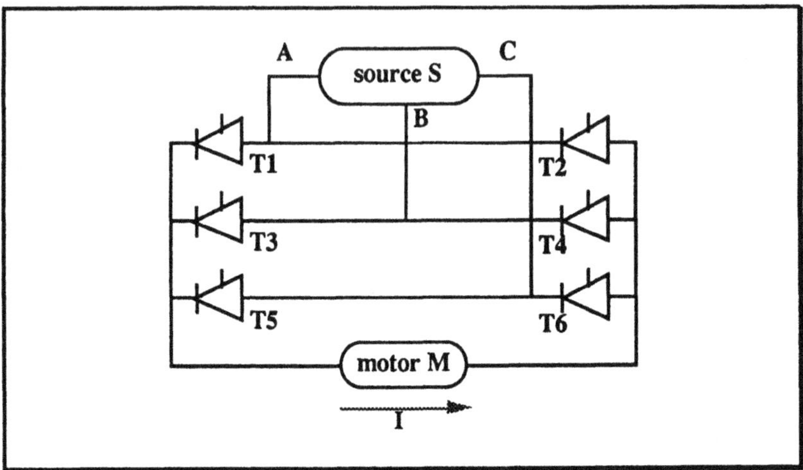

Fig. 69 – A rectifier circuit with thyristors

only (a part of) the positive voltage of the source becomes effective. An external control unit fires the thyristors in such a way that at any given time two of them establish a connection between two poles with a positive voltage drop. As a result, the motor is supplied with rectified current I. Second, the thyristors are used for controlling the current. The current supply obviously depends on the times when the thyristors are fired. If the gate pulse arrives at a time $t_0' < t_0$, more of the positive voltage is exploited and the average current is increased. Conversely, firing the thyristor later reduces the average current.

Under normal operating conditions the effective current through M oscillates around the average with six times the frequency of a phase (fig. 70 (a)). In SHERLOCK terminology, this situation corresponds to the candidate $\{OK(T_1), ..., OK(T_6)\}$. However, thyristors are known to fail in several possible ways for which we can construct fault models:

- B(T): the thyristor is always in the blocking state;
- C(T): the thyristor is always in the conducting state;
- ...

All faults for which no explicit models exist are lumped together in U(T). If one or more of the thyristors are in an abnormal mode, this is reflected in the effective current in a characteristic way.

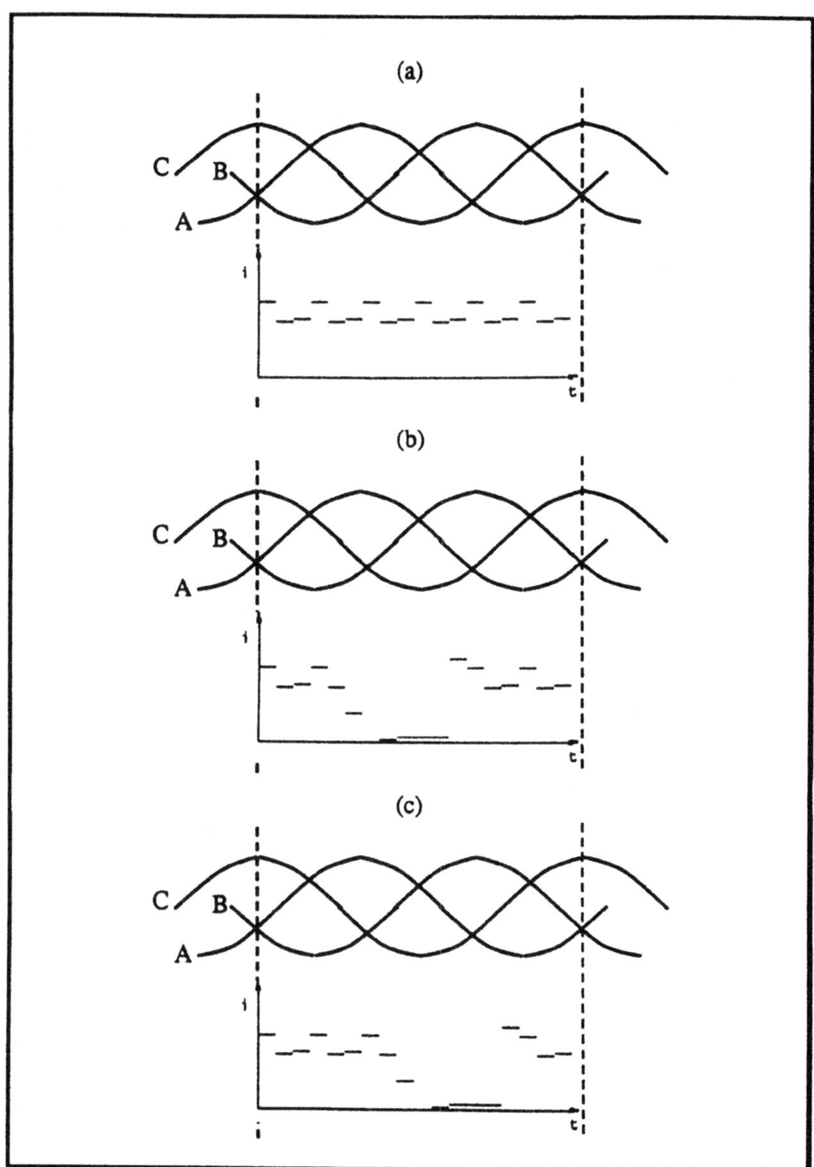

Fig. 70 – Effective current diagrams for (a) normal operating conditions, (b) blocking thyristor T_1, (c) blocking thyristor T_6. The current diagrams have been reproduced directly from the simulation results reported in [Decker89], whereas the phase voltages have been added for reference.

As an example, fig. 70 (b) and (c) show the effective current diagrams for the candidates $K_1 = \{B(T_1), OK(T_2), ..., OK(T_6)\}$ and $K_6 = \{OK(T_1), ..., OK(T_5),$

$B(T_6)\}$. Looking at the two curves, the problem of discriminating between K_1 and K_2 becomes obvious: the two current curves are absolutely identical except for their temporal relation to the three phases. No static measurement of the current alone can possibly reveal which of the candidates represents the real situation, instead discrimination must be based on a generalized measurement of the effective current I *and* the phases A, B, and C. Notice that the working assumptions of dynamic SHERLOCK are satisfied in this case:

- General background knowledge in electrical engineering is sufficient to consider "looking at voltages and currents" a potentially useful experiment.
- The behavior of the device over time in terms of A, B, C, and I can be predicted for each candidate.
- Based on the predictions we can use the augmented discrimination algorithm from section 4.2.3.2.

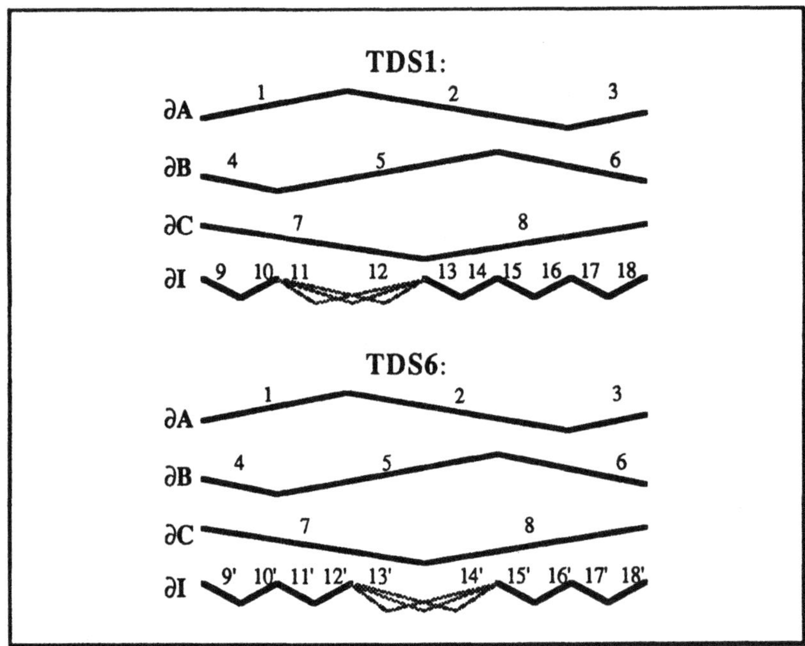

Fig. 71 – Graphical representation of the TDS descriptions TDS1 and TDS6. Upward-slanting line segments correspond to episodes with value ↑, downward-slanting segments to episodes with value ↓. Episodes 11, 12, 13' and 14' have disjunctive temporal relations with the rest of the episodes.

Suppose that we have reached the following stage during diagnosis. The current high-probability candidates are $K_i = \{OK(T_1), ..., OK(T_{i-1}), B(T_i), OK(T_{i+1}), ..., OK(T_6)\}$, $i = 1,...,6$, with equal probabilities $p(K_1) = ... = p(K_6) =: p$. There are no static measurements with non-zero expected information gain, but via simulation we have computed $Beh(K_i, S_0, \{A,B,C,I\})$, $i = 1,...,6$, where S_0 is an arbitrary state in the repeating cycle formed by the three phases. For simplicity we perform the following steps for only two candidates, K_1 and K_6. We abstract from the real-valued quantities A, B, C, and I by considering only the signs of their derivatives (denoted by ∂A, ∂B, ∂C, ∂I with values ↑ and ↓) and obtain $Beh(K_1, S_0, \{\partial A,\partial B,\partial C,\partial I\}) = \{TDS1\}$ and $Beh(K_6, S_0, \{\partial A,\partial B,\partial C,\partial I\}) = \{TDS6\}$ as shown in fig. 71. The two TDS descriptions are largely identical, the only differences being the temporal relations between episodes 11–14 resp. 11'–14' and the rest of the episodes.

We now execute PREPARE_AND_EVAL_MSMT for K_1 and K_6. Neither K_1 nor K_6 contains a component in a U-mode, so `S*` is set to {TDS1, TDS6} in step (1). In (5) and (6) we initialize one instantiation record for each TDS description and perform algorithm SUGGEST for them. As the beginnings of TDS1 and TDS6 are identical, SUGGEST suggests the same quantities to be measured in both cases, ∂C and ∂I, which we add to Q^*. We also record in `V(∂C)` and `V(∂I)` that in both cases we expect to measure ∂C = ↓ and ∂I = ↓ (because we expect to observe episodes 7 and 9 (9')). None of the TDS descriptions has been fully matched yet, so we continue with step (12) imposing an arbitrary order on the two proposed quantities, say ⟨∂C, ∂I⟩. We then add $\langle \partial C, t_1\rangle$ and $\langle \partial I, t_2\rangle$ to the initially empty measurement plan `MP({∂A, ∂B, ∂C, ∂I})` where $t_1 < t_2$ are two time points in the time interval during which the actual measurements will take place. In step (15) `V(∂C) × V(∂I)` contains only one element, ⟨↓, ↓⟩, so we execute MATCH for the two instantiations and the measurements $\langle \partial C, t_1, \downarrow\rangle$ and $\langle \partial I, t_2, \downarrow\rangle$. In both cases the matches succeed and we reach step (24) where we replace `T` with `NewT` containing the instantiation records ⟨TDS1, G'(TDS1), $\{\langle \partial C, t_1, \downarrow\rangle, \langle \partial I, t_2, \downarrow\rangle\}$⟩ and ⟨TDS6, G'(TDS6), $\{\langle \partial C, t_1, \downarrow\rangle, \langle \partial I, t_2, \downarrow\rangle\}$⟩. In similar fashion the next measurements $\langle \partial A, t_3, \uparrow\rangle$, $\langle \partial B, t_4, \downarrow\rangle$, $\langle \partial I, t_5, \uparrow\rangle$, $\langle \partial B, t_6, \uparrow\rangle$, $\langle \partial I, t_7, \downarrow\rangle$, $\langle \partial A, t_8, \uparrow\rangle$ are added. Now the two TDS descriptions diverge: based on TDS6's instantiation record SUGGEST suggests ∂I (with expected value ↑) whereas TDS1 produces the additional suggestion ∂A (with expected value ↓). Hence, $Q^* = \{\partial A, \partial I\}$, $V(\partial A) = \{\uparrow, \downarrow\}$ and $V(\partial I) = \{\uparrow\}$. We add $\langle \partial A, t_9\rangle$ and $\langle \partial I, t_{10}\rangle$ to the measurement plan and execute MATCH *twice* for each instantiation: once for the measurements $\langle \partial A, t_9, \uparrow\rangle$, $\langle \partial I, t_{10}, \uparrow\rangle$ and once for $\langle \partial A, t_9, \downarrow\rangle$, $\langle \partial I, t_{10}, \uparrow\rangle$. The match of $\langle \partial A, t_9, \downarrow\rangle$, $\langle \partial I, t_{10}, \uparrow\rangle$ against TDS6's instantiation fails, whereas the other three succeed, leading to three new instantiations:

$\langle$TDS1, G'(TDS1), {..., $\langle \partial A, t_9, \uparrow\rangle, \langle \partial I, t_{10}, \uparrow\rangle\}\rangle$,
$\langle$TDS1, G"(TDS1), {..., $\langle \partial A, t_9, \downarrow\rangle, \langle \partial I, t_{10}, \uparrow\rangle\}\rangle$,
$\langle$TDS6, G"'(TDS6), {..., $\langle \partial A, t_9, \uparrow\rangle, \langle \partial I, t_{10}, \uparrow\rangle\}\rangle$.

From now on these three instantiations are processed quasi-concurrently. A few measurements later again new instantiations are created – this time for TDS6. However, due to the arbitrary tie-breaking between quantities suggested in parallel, the new instantiations are eliminated after some more measurements. The procedure terminates when – after a total of 26 individual measurements – the three descendents of the instantiations shown above have simultaneously been fully matched. In step (25) we then extract the accumulated measurement sequences from the instantiations (two for TDS1 and one for TDS6). These measurement sequences form the sets `Seq(K`$_i$`, S`$_0$`, MP({∂A,∂B,∂C,∂I}))`.

It turns out that these sets are disjoint so that, no matter what the particular result is, only one candidate is left over after the generalized measurement $\langle S_0$, MP({∂A,∂B,∂C,∂I})$\rangle$. The same applies when all six candidates are considered (only the measurement plan contains more measurements). Thus the expected a posteriori entropy is zero which is better than the current entropy and, consequently, the generalized measurement is selected.

4.2.3.4 Evaluation: When to Use Generalized Measurements

Under which circumstances can it be of advantage to use generalized measurements instead of individual measurements à la static SHERLOCK? Clearly, the effort spent on predicting dynamic behaviors, preparing a generalized measurement, predicting the results for each candidate and performing several measuring actions instead of only one is significantly greater in the dynamic case. As we have pointed out several times, the single great advantage of generalized measurements is their ability to discriminate between candidates that are indistinguishable by static measurements. A rational strategy would therefore save non-static measurements until they are really needed and thus split the diagnostic process into two distinct phases. During phase 1 static models are used for candidate generation and the static SHERLOCK procedure is employed to reduce the set of candidates as far as possible. This is no contradiction in terms: Any device - dynamic or not - can be modeled statically on a sufficiently high level of abstraction. Devices differ only in how far one can get in hypothesis discrimination before dynamic aspects become crucial. Phase 1 can terminate for several reasons:

- The final diagnosis has been found.

- The remaining candidates can be distinguished only by TDSs, i.e. the remaining static measurements show no expected decrease in entropy.

- There are static measurements left which could further reduce the candidate set, but for some reason (test cost, accessibility, equipment, ...) they cannot be applied. Generalized measurements may be used as substitutes, if *their* individual measurements are practical.

Only in the second and third cases we would switch to dynamic models and use dynamic SHERLOCK in phase 2 to determine the final diagnosis. This division of labor has an important pragmatic advantage. Not all faults are so complex that they require generalized measurements. This implies that the dynamic models, which are considerably harder to construct than static models, need not necessarily cover the whole device, but only those areas in which dynamics actually play a role. Smaller models not only simplify the knowledge acquisition task but improve also the performance of the simulator used to predict the dynamic behavior of the device.

5 Conclusion

5.1 Status of the Implementation

All of the algorithms described in chapter 3 and section 4.1 have been fully implemented and tested on examples. The first prototype of the temporal matching algorithm was implemented by myself in Temporal Prolog, an extension of Prolog described in [Hrycej87] which supports Allen's interval calculus. Later versions and the extension to duration bounds were implemented in Smalltalk-80, the language underlying the MOLTKE system, by Hans Lamberti [Lamberti88], Johannes Stein [Stein88], and Klaus Becker [BeckerK89].

The architecture for dynamic SHERLOCK proposed in section 4.2 has not yet led to a complete working system. However, parts of the system are currently being implemented as the basis for the more ambitious universal solution outlined in section 5.3.

5.2 Evaluation

Based on our experience with the temporal matching algorithm, its extensions and the embedding into the MOLTKE system we can now evaluate whether and to which extent the original goals of the research have been achieved.

As we have explained already in the introduction, the research was motivated in the first place by the need for a systematic treatment of TDSs within MOLTKE. To be useful and practical in this context a solution had to be both reasonably efficient and sufficiently expressive. Our experiences indicate that the TDS description language and the temporal matching algorithm in its extended version meet these requirements.

However, it must not be overlooked that some of the working assumptions and design decisions are pragmatic answers to otherwise open questions (e.g. the choice of granularity bounds as one form of representing persistence information). While justified in MOLTKE's domain, these assumptions need to be carefully verified for other potential application domains. Nevertheless, we expect most of our basic axioms to be general enough to be satisfied in other domains and experiments with TDSs for different kinds of devices support this claim.

For temporal matching to be a universal tool it is furthermore desirable that the technique is independent of the particular diagnostic paradigm. In chapter 4 we have demonstrated that our approach can be utilized at least in the two most popular paradigms: associative/heuristic and model-based diagnosis.

The results about convex relations (section 3.2.5) are of an altogether different nature. Although the search for a tractable subset of the full relation algebra was guided by the diagnostic task, the results are more general. After all, dynamic behavior has to be represented in AI systems for various tasks (e.g. monitoring systems, qualitative simulation, etc.) and they all benefit from the attractive computational properties of convex relations, as long as the subset is sufficiently expressive for their purposes. The novel characterization of convex relations can be of great help in determining their applicability.

5.3 Open Problems and Future Work

Some of the lines along which research on TDSs can be continued follow canonically from the limitations of our approach which we have mentioned at various points in our presentation.

Historically, our perspective on TDSs has been strongly influenced by their role in MOLTKE's application domain. Although we have strived to avoid overly domain-specific design decisions, the claim to generality will ultimately have to be borne out by experiments in other domains. Work in this direction has been started in the area of thermal engines (e.g. refrigerators) and the preliminary results are encouraging.

But transfer to new domains need not be limited to fault diagnosis. As a basic technique, we can imagine that temporal matching can be applied in a different context which, too, requires reasoning about the dynamic behavior of technical systems. For example, in reactive planning, feedback is gained by monitoring the plan execution and comparing the observations to the predicted behavior. If the predictions could be described in the form of a TDS description, temporal matching might be an instrument in designing efficient monitoring schedules.

Other open problems are more of a technical than conceptual nature. One of the most urgent problems in this category is the development of a more user-friendly interface for specifying TDS descriptions. As mentioned in section 4.1.2.1, a graphical user interface would be a great improvement over the current external syntax, both in terms of ease of use and of security against mistakes in specifications. We would like to stress that – not unlike the spin-off results about convex relations – such a development, while only of technical importance for our system, might bring about new results in graph theory and related areas which are interesting in their own right. [Stein88] contains some initial ideas about this topic.

Finally, there is one particularly fascinating question that we deliberately sidestepped in section 4.2.3: Where do TDSs come from? An answer may be found in the model-based approach to technical diagnosis. Earlier we explained that in dynamic SHERLOCK due to the sheer number of potential generalized measurements we cannot afford to enumerate them all and rank them according to their expected information gain. Our pragmatic solution at the time was to exploit heuristic information in the technical area to filter out only promising generalized measurements. Nevertheless we would prefer a more general solution which determines an optimal (or near-optimal) generalized measurement solely on the basis of the available models. We are currently studying possible techniques for this solution; one such method is outlined below.

The key idea is to reverse the whole process of measurement selection: we start with the desired effect (i.e. we decide which candidate(s) we would like to eliminate next) and construct a specific generalized measurement that will accomplish the task. In full generality this problem is at least as hard as the former, but the task becomes easier, if we restrict ourselves to certain patterns of candidate elimination which we call the *effect type* of the generalized measurement. After we have chosen a particular effect of the type we have at our disposal, we must synthesize a generalized measurement and predict the candidate behaviors for it.

An example of an effect type can be characterized as follows: Suppose that $K^* = \{K_1, \ldots, K_m\}$ is the current set of candidates, that $p(K_l)$ are their probabilities and that H is the current entropy. Pick two candidates $K_i, K_j \in K^*$, $i \neq j$, without U-modes and imagine that there were a generalized measurement that is known to establish the following situation:

- If K_j is the actual diagnosis, then K_i is eliminated;
- if K_i is the actual diagnosis, then K_j is eliminated;
- if neither K_i nor K_j is the actual diagnosis, then none, one, or both of K_i and K_j may be eliminated.

What would be the consequence? Judging conservatively, none of the other candidates are affected in this step so that the new set of candidates is either $K_i^* = K^* \setminus \{K_i\}$ or K_j^*

$= K^* \setminus \{K_j\}$ or $K^*_{ij} = K^* \setminus \{K_i, K_j\}$ or K^* itself. Using a simplified version of equation (**) we can estimate the information gain of the hypothetical measurement. Repeating this step for each pair i,j we determine the pair i_{opt}, j_{opt} for which the expected information gain is maximal.

Of course, the hard part of the task is constructing a generalized measurement which will discriminate between candidates $K_{i_{opt}}$ and $K_{j_{opt}}$. Even conceding that there may always be candidates which are in principle indistinguishable and that we make the assumption only in cases where discrimination is *at all* possible the task is far from trivial. Already the introductory car engine example illustrates the enormous search space for generalized measurements when arbitrarily complex external actions (e.g. changes in the topology of the device) are allowed. Here we run into the problems of general action planning in a reactive environment. In addition, the current state of the art in model-based reasoning about dynamic systems is itself characterized by severe complexity problems so that a universal solution does not appear practical. However, if we limit ourselves to a very simple type of actions (selecting an initial state), the task bears a strong resemblance to automated test generation. Consequently our current research is aimed at generalizing a well-known test generation algorithm, the so-called D-algorithm [Roth67] for our purpose.

5.4 Final Words

If we finally step back from the work reported in this thesis, we are bound to reach the inevitable conclusion that is quintessential of scientific work in general: for each problem solved, many new questions have been raised. We hope that we have succeeded in conveying to the reader a piece of the fascination of both.

Appendix

A. BNF Definition of the External Syntax for TDS Descriptions in MOLTKE

<TDS_description> ::= (TNET <epi_decl>+ <qual_constr>+ <quant_constr>*)

<epi_decl> ::= (<quantity> <op> <value> <episode_id>)

<qual_constr> ::= (<episode_id> (<prim_rel>+) <episode_id>)

<quant_constr> ::= (<bound_type> <episode_id> <episode_id>)

<op> ::= = | > | < | ≥ | ≤

<quantity> ::= *identifier*

<value> ::= *identifier*

<episode_id> ::= *identifier*

<prim_rel> ::= e | b | bi | o | oi | m | mi | s | si | f | fi | d | di

<bound_type> ::= LOWER | UPPER

B. List of Definitions

DEFINITION 1: time net
DEFINITION 2: singleton time net, unconstrained time net
DEFINITION 3: models of a time net
DEFINITION 4: combinable functions, combination function
DEFINITION 5: projection of models
DEFINITION 6: restriction of models
DEFINITION 7: interval-convex
DEFINITION 8: convex relations
DEFINITION 9: P-definable interval relations
DEFINITION 10: TDS description
DEFINITION 11: TDS time net
DEFINITION 12: instance of a TDS description, basis interval
DEFINITION 13: occurs (for instances of a TDS description)
DEFINITION 14: measurement, observation
DEFINITION 15: measurement sequence
DEFINITION 16: compatible (measurement sequence with TDS instance)
DEFINITION 17: granularity of a TDS instance
DEFINITION 18: granularity of a measurement sequence
DEFINITION 19: weakly determines
DEFINITION 20: matches (measurement sequence against TDS description)
DEFINITION 21: TDS descriptions with duration bounds
DEFINITION 22: matches_with_durations
DEFINITION 23: behaviors for a candidate K, a set of quantities Q, starting in the initial state IS
DEFINITION 24: measurement plan
DEFINITION 25: generalized measurement
DEFINITION 26: possible measurement sequences
DEFINITION 27: compatible measurement sequences

C. List of Figures

Related Work

Temporal Matching

Incorporating TDSs into Existing Diagnostic Paradigms

D. References

[Allen/Hayes85] J. F. Allen, P. F. Hayes: A Common Sense Theory of Time, in: Proc. 9th IJCAI, 1985

[Allen83] J. F. Allen: Maintaining Knowledge about Temporal Intervals, in: Comm. ACM **26**(11), November 1983

[Althoff et al. 88] K. Althoff, K. Nökel, R.Rehbold, M.M. Richter: A Sophisticated Expert System for the Diagnosis of a CNC-Machining Center, in: Zeitschrift für Operations Research, vol. **32**, 1988

[Althoff et al. 89a] K. Althoff, S. Kockskämper, F. Maurer, M. Stadler, S. Weß: Ein System zur fallbasierten Wissensverarbeitung in technischen Diagnosesituationen, in: Proc. ÖGAI Annual Conference, Igls / Innsbruck, 1989

[Althoff et al. 89b] K. Althoff, A. de la Ossa, F. Maurer, M. Stadler, S. Weß: Adaptive Learning in the Domain of Technical Diagnosis, in: Proc. FAW Workshop on Adaptive Learning, Ulm, 1989

[BeckerK89] K. Becker: Temporales Matching unter Berücksichtigung von Intervallängenrestriktionen, term project, Universität Kaiserslautern, 1989

[BeckerU89] U. Becker: Modellgestützte Testsimulation mit Zeitpropagierung für wissensbasierte Fehlerdiagnose von elektronischen Schaltungen, diploma thesis, Universität Kaiserslautern, 1989

[Box/Jenkins70] G.E.P. Box, G.M. Jenkins: Time Series Analysis, Forecasting and Control, Holden-Day, San Francisco, 1970

[Bruce72] B.C. Bruce: A Model for Temporal Reference and its Application in a Question Answering Program, in: Artificial Intelligence **3** (1972)

[de Kleer/Williams86] J. de Kleer, B.C. Williams: Reasoning about Multiple Faults, in: Proc. AAAI-86

[de Kleer/Williams87] J. de Kleer, B.C. Williams: Diagnosing Multiple Faults, in: Artificial Intelligence **32** (1987), also in: M.L. Ginsberg (ed.): Readings in Nonmonotonic Reasoning, Morgan Kaufmann, 1987 [de Kleer/Williams89] J. de Kleer, B.C. Williams: Diagnosis as Identifying Consistent Modes of Behavior, in: Proc. 11th IJCAI, Detroit, 1989

[de Kleer89] J. de Kleer: Entropy without Probabilities, in: Proc. International Workshop on Model-Based Diagnosis, Paris, 1989

[Decker87] R. Decker: Zeitliches Schließen in Constraint-Systemen, Siemens AG, Report INF2 ARM-4-87, München, 1987

[Decker89] R. Decker: Qualitative Simulation des zeitlichen Verhaltens von Thyristorbrückengleichrichterschaltungen, Technical Report, Siemens AG, München 1989

[deKleer/Brown84] J. deKleer, J. S. Brown: A Qualitative Physics Based on Confluences, in: D. G. Bobrow (ed.): Qualitative Reasoning about Physical Systems, Amsterdam 1984

[Doyle et al. 89] R.J. Doyle, S.M. Sellers, D.J. Atkinson: A Focused, Context-Sensitive Approach to Monitoring, in: Proc. 11th IJCAI, Detroit, 1989

[Dvorak/Kuipers89] D. Dvorak, B. Kuipers: Model-Based Monitoring of Dynamic Systems, in: Proc. 11th IJCAI, Detroit, 1989

[Fagan84] L. Fagan, J. Kunz, E. Feigenbaum, J. Osborn: Extensions to the Rule-Based Formalism for Monitoring Tasks, in: B. Buchanan, E. Shortliffe (eds.): Rule-Based Expert Systems, Addison-Wesley, 1984

[Forbus83] K. D. Forbus: Measurement Interpretation in Qualitative Process Theory, in: Proc. 8th IJCAI, 1983

[Forbus84] K. D. Forbus: Qualitative Process Theory, in: D. G. Bobrow (ed.): Qualitative Reasoning about Physical Systems, Amsterdam 1984

[Forbus86] K. D. Forbus: Interpreting Measurements of Physical Systems, in: Proc. AAAI-86

[Guckenbiehl89] T. Guckenbiehl: Ein Ansatz zur Erweiterung der Episodenpropagierung, in: Proc. ÖGAI-Jahrestagung 1989, Springer, 1989

[Güsgen/Fidelak88] H.W. Güsgen, M. Fidelak: Constraint Satisfaction with Timed Values, manuscript, GMD, 1988

[Hamscher/Davis84] W. Hamscher, R. Davis: Diagnosing Circuits with State: an Inherently Underconstrained Problem, in: Proc. AAAI-84

[Hrycej88] T. Hrycej: Temporal Prolog, in: Proc. ECAI-88

[Kautz88] H. Kautz: *private communication*

[Kockskämper89] S. Kockskämper: Diskussion möglicher Wissensrepräsentations- und Inferenzmechanismen zur Fehlerdiagnose eines CNC-Bearbeitungszentrums und deren Implementierung als Expertensystem in SMALLTALK-80, diploma thesis, Universität Kaiserslautern, 1989

[Kuipers86] B. Kuipers: Qualitative Simulation, in: Artificial Intelligence, vol. **29** (3), September 1986

[Lamberti88] H. Lamberti: Ein temporaler Matching-Algorithmus für dynamische Fehlersituationen, term project, Universität Kaiserslautern, 1988

[Leitch, Wiegand89] R. Leitch, M. Wiegand: Temporal Issues in Qualitative Reasoning, in: Proc. ÖGAI-Jahrestagung 1989, Springer, 1989

[McDermott82] D. McDermott: A Temporal Logic for Reasoning about Processes and Plans, in: Cognitive Science **6** (1982)

[Moissiadis90] C. Moissisadis: Repräsentation von Aufbauplänen technischer Geräte und deren Nutzung zur Herleitung kausaler Regeln und modellbasierter Diagnose, diploma thesis, Universität Kaiserslautern, 1990

[Neumann84] B. Neumann: Natural Language Description of Time-Varying Scenes, Report no. 105, FB Informatik, Universität Hamburg, 1984

[Nökel/Lamberti90] K. Nökel, H. Lamberti: Temporally Distributed Symptoms in the MOLTKE System for Technical Diagnosis, submitted to Journal of Artificial Intelligence in Engineering

[Nökel89a] K. Nökel: Convex Relations between Time Intervals, in: Proc. ÖGAI-Jahrestagung, Igls / Innsbruck, 1989

[Nökel89b] K. Nökel: Temporal Matching: Recognizing Dynamic Situations from Discrete Measurements, in: Proc 11th IJCAI, Detroit, 1989

[Patil81] R.S. Patil: Causal Representation of Patient Illness for Electrolyte and Acid-Base Diagnosis, PhD thesis, MIT, 1981

[Pawlak82] Z. Pawlak: Rough Sets, in: Journal of Information and Computer Sciences **11** (1982)

[Puppe87] F. Puppe: Diagnostisches Problemlösen mit Expertensystemen, Informatik-Fachberichte 148, Springer, 1987

[Puppe88] F. Puppe: Einführung in Expertensysteme, Studienreihe Informatik, Springer, 1988

[Rehbold89] R. Rehbold: Model-Based Knowledge Acquisition from Structure Descriptions in a Technical Diagnosis Domain, in: Proc. 9th International Workshop on Expert Systems and their Applications, Special Conference on Second Generation Expert Systems, Avignon, 1989

[Richter89] M.M. Richter: Prinzipien der künstlichen Intelligenz, Teubner 1989

[Rist et al.87] T. Rist, G. Herzog, E. Andre: Ereignismodellierung zur inkrementellen High-level Bildfolgenanalyse, VITRA-report no. 19, KI-Labor, Universität des Saarlandes, 1987

[Rit88] J.-F. Rit: Modélisation et Propagation de Contraintes Temporelles Pour la Planification, PhD thesis, Grenoble 1988

[Roth et al.67] J.P. Roth, W.G. Bouricius, P.R.Schneider: Programmed Algorithms to Compute Tests to Detect and Distinguish Between Failures in Logic Circuits, in: IEEE Transactions on Electronic Computers, vol. EC-16, no. 5, 1967

[Schmiedel88] A. Schmiedel: Temporal Constraint Networks, KIT-Report 69, Tech. University Berlin, project group KIT, 1988

[Schuch89] A. Schuch: Ein Verfahren zur Meßpunktauswahl in der technischen Diagnose, term project, Universität Kaiserslautern, 1989

[Shirley88] M.H. Shirley: Generating Circuit Tests by Exploiting Designed Behavior, PhD thesis, MIT AI Lab TR 1099, 1988

[Shoham88] Y. Shoham: Reasoning About Change, MIT Press, Cambridge, 1988

[Shortliffe76] E.H. Shortliffe: MYCIN: Computer-Based Medical Consultations, Elsevier, 1976 (based on Shortliffe's PhD thesis, Stanford University, 1974)

[Stallman/Sussman79] R.M. Stallman, G.J. Sussman: Problem Solving About Electrical Circuits, in: P.H. Winston, R.H. Brown (eds.): Artificial Intelligence: An MIT Perspective, vol. 1, MIT Press, 1979

[Stein88] J. Stein: Entwicklung und Implementierung eines Verfahrens zur Erkennung dynamischer Fehlersituationen, term project, Universität Kaiserslautern, 1988

[Struss/Dressler89] P. Struss, O. Dressler: "Physical Negation" – Integrating Fault Models into the General Diagnostic Engine, in: Proc. 11th IJCAI, Detroit, 1989

[Struss89] P. Struss: Diagnosis as a Process, extended abstract, in: Proc. International Workshop on Model-Based Diagnosis, Paris, 1989

[Struss90] P. Struss: Model-Based Diagnosis – Recent Advances and Perspectives, Technical Report, Siemens AG, München, 1990

[Tsotsos85] J. K. Tsotsos: Knowledge organization and its role in representation and interpretation for time-varying data: the ALVEN system, in: Computational Intelligence **1** (1985)

[Valdes-Perez87] R.E. Valdés-Pérez: The Satisfiability of Temporal Constraint Networks, in: Proc. AAAI-87

[vanBeek89] P. van Beek: Approximation Algorithms for Temporal Reasoning, in: Proc. 11th IJCAI, Detroit, 1989

[Vilain et al. 90] M. Vilain, H. Kautz, P. van Beek: Constraint Propagation Algorithms for Temporal Reasoning – A Revised Report, in: D.S. Weld, J. de Kleer (eds.): Readings in Qualitative Reasoning about Physical Systems, Morgan Kaufmann, 1990

[Vilain/Kautz86] M. Vilain, H. Kautz: Constraint Propagation Algorithms for Temporal Reasoning, Proc. AAAI-86

[Voß87] H. Voß: Representing and Analyzing Causal, Temporal, and Hierarchical Relations of Devices, SEKI-Report SR-87-17, Universität Kaiserslautern

[Wetter84] T. Wetter: Ein modallogisch beschriebenes Expertensystem, ausgeführt am Beispiel von Ohrenerkrankungen, PhD thesis, RWTH Aachen, 1984

[Williams86] B.C. Williams: Doing Time - Putting Qualitative Reasoning on Firmer Ground, in: Proc. AAAI-86

[Zadeh79] L.A. Zadeh: A Theory of Approximate Reasoning, in: Machine Intelligence **9** (1979)

E. Index of Informally Introduced Concepts

Lecture Notes in Artificial Intelligence (LNAI)

Vol. 345: R.T. Nossum (Ed.), Advanced Topics in Artificial Intelligence. VII, 233 pages. 1988.

Vol. 346: M. Reinfrank, J. de Kleer, M.L. Ginsberg, E. Sandewall (Eds.), Non-Monotonic Reasoning. Proceedings, 1988. XIV, 237 pages. 1989.

Vol. 347: K. Morik (Ed.), Knowledge Representation and Organization in Machine Learning. XV, 319 pages. 1989.

Vol. 353: S. Hölldobler, Foundations of Equational Logic Programming. X, 250 pages. 1989.

Vol. 383: K. Furukawa, H. Tanaka, T. Fujisaki (Eds.), Logic Programming '88. Proceedings, 1988. IX, 251 pages. 1989.

Vol. 390: J.P. Martins, E.M. Morgado (Eds.), EPIA 89. Proceedings. 1989. XII, 400 pages. 1989.

Vol. 395: M. Schmidt-Schauß, Computational Aspects of an Order-Sorted Logic with Term Declarations. VIII, 171 pages. 1989.

Vol. 397: K.P. Jantke (Ed.), Analogical and Inductive Inference. Proceedings, 1989. IX, 338 pages. 1989.

Vol. 406: C.J. Barter, M.J. Brooks (Eds.), AI '88. Proceedings, 1988. VIII, 463 pages. 1990.

Vol. 418: K.H. Bläsius, U. Hedtstück, C.-R. Rollinger (Eds.), Sorts and Types in Artificial Intelligence. Proceedings, 1989. VIII, 307 pages. 1990.

Vol. 419: K. Weichselberger, S. Pöhlmann, A Methodology for Uncertainty in Knowledge-Based Systems. VIII, 132 pages. 1990.

Vol. 422: B. Nebel, Reasoning and Revision in Hybrid Representation Systems. XII, 270 pages. 1990.

Vol. 437: D. Kumar (Ed.), Current Trends in SNePS – Semantic Network Processing System. Proceedings, 1989. VII, 162 pages. 1990.

Vol. 444: S. Ramani, R. Chandrasekar, K.S.R. Anjaneyulu (Eds.), Knowledge Based Computer Systems. Proceedings, 1989. X, 546 pages. 1990.

Vol. 446: L. Plümer, Termination Proofs for Logic Programs. VIII, 142 pages. 1990.

Vol. 449: M.E. Stickel (Ed.), 10th International Conference on Automated Deduction. Proceedings, 1990. XVI, 688 pages. 1990.

Vol. 451: V. Marík, O. Stepánková, Z. Zdráhal (Eds.), Artificial Intelligence in Higher Education. Proceedings, 1989. IX, 247 pages. 1990.

Vol. 459: R. Studer (Ed.), Natural Language and Logic. Proceedings, 1989. VII, 252 pages. 1990.

Vol. 462: G. Gottlob, W. Nejdl (Eds.), Expert Systems in Engineering. Proceedings, 1990. IX, 260 pages. 1990.

Vol. 465: A. Fuhrmann, M. Morreau (Eds.), The Logic of Theory Change. Proceedings, 1989. X, 334 pages. 1991.

Vol. 475: P. Schroeder-Heister (Ed.), Extensions of Logic Programming. Proceedings, 1989. VIII, 364 pages. 1991.

Vol. 476: M. Filgueiras, L. Damas, N. Moreira, A.P. Tomás (Eds.), Natural Language Processing. Proceedings, 1990. VII, 253 pages. 1991.

Vol. 478: J. van Eijck (Ed.), Logics in AI. Proceedings. 1990. IX, 562 pages. 1991.

Vol. 481: E. Lang, K.-U. Carstensen, G. Simmons, Modelling Spatial Knowledge on a Linguistic Basis. IX, 138 pages. 1991.

Vol. 482: Y. Kodratoff (Ed.), Machine Learning – EWSL-91. Proceedings, 1991. XI, 537 pages. 1991.

Vol. 513: N. M. Mattos, An Approach to Knowledge Base Management. IX, 247 pages. 1991.

Vol. 515: J. P. Martins, M. Reinfrank (Eds.), Truth Maintenance Systems. Proceedings, 1990. VII, 177 pages. 1991.

Vol. 517: K. Nökel, Temporally Distributed Symptoms in Technical Diagnosis. IX, 164 pages. 1991.

Vol. 518: J. G. Williams, Instantiation Theory. VIII, 133 pages. 1991.

Lecture Notes in Computer Science

Vol. 483: G. Rozenberg (Ed.), Advances In Petri Nets 1990. VI, 515 pages. 1991.

Vol. 484: R. H. Möhring (Ed.), Graph-Theoretic Concepts In Computer Science. Proceedings, 1990. IX, 360 pages. 1991.

Vol. 485: K. Furukawa, H. Tanaka, T. Fullsaki (Eds.), Logic Programming '89. Proceedings, 1989. IX, 183 pages. 1991. (Subseries LNAI).

Vol. 486: J. van Leeuwen, N. Santoro (Eds.), Distributed Algorithms. Proceedings, 1990. VI, 433 pages. 1991.

Vol. 487: A. Bode (Ed.), Distributed Memory Computing. Proceedings, 1991. XI, 506 pages. 1991.

Vol. 488: R. V. Book (Ed.), Rewriting Techniques and Applications. Proceedings, 1991. VII, 458 pages. 1991.

Vol. 489: J. W. de Bakker, W. P. de Roever, G. Rozenberg (Eds.), Foundations of Object-Oriented Languages. Proceedings, 1990. VIII, 442 pages. 1991.

Vol. 490: J. A. Bergstra, L. M. G. Feljs (Eds.), Algebraic Methods 11: Theory, Tools and Applicatlons. VI, 434 pages. 1991.

Vol. 491: A. Yonezawa, T. Ito (Eds.), Concurrency: Theory, Language, and Architecture. Proceedings, 1989. VIII, 339 pages. 1991.

Vol. 492: D. Sriram, R. Logcher, S. Fukuda (Eds.), Computer-Aided Cooperative Product Development. Proceedings, 1989 VII, 630 pages. 1991.

Vol. 493: S. Abramsky, T. S. E. Maibaum (Eds.), TAPSOFT '91. Volume 1. Proceedings, 1991. VIII, 455 pages. 1991.

Vol. 494: S. Abramsky, T. S. E. Maibaum (Eds.), TAPSOFT '91. Volume 2. Proceedings, 1991. VIII, 482 pages. 1991.

Vol. 495: 9. Thalheim, J. Demetrovics, H.-D. Gerhardt (Eds.), MFDBS '91. Proceedings, 1991. VI, 395 pages. 1991.

Vol. 496: H.-P. Schwefel, R. Männer (Eds.), Parallel Problem Solving from Nature. Proceedings, 1991. XI, 485 pages. 1991.

Vol. 497: F. Dehne, F. Fiala. W.W. Koczkodaj (Eds.), Advances in Computing and Intormation - ICCI '91 Proceedings, 1991. VIII, 745 pages. 1991.

Vol. 498: R. Andersen, J. A. Bubenko jr., A. Sølvberg (Eds.), Advanced Information Systems Engineering. Proceedings, 1991. VI, 579 pages. 1991.

Vol. 499: D. Christodoulakis (Ed.), Ada: The Choice for '92. Proceedings, 1991. VI, 411 pages. 1991.

Vol. 500: M. Held, On the Computational Geometry of Pocket Machining. XII, 179 pages. 1991.

Vol. 501: M. Bidoit, H.-J. Kreowski, P. Lescanne, F. Orejas, D. Sannella (Eds.), Algebraic System Specification and Development. VIII, 98 pages. 1991.

Vol. 502: J. Bārzdiņš, D. Bjørner (Eds.), Baltic Computer Science. X, 619 pages. 1991.

Vol. 503: P. America (Ed.), Parallel Database Systems. Proceedings, 1990. VIII, 433 pages. 1991.

Vol. 504: J. W. Schmidt, A. A. Stogny (Eds.), Next Generation Information System Technology. Proceedings, 1990. IX, 450 pages. 1991.

Vol. 505: E. H. L. Aarts, J. van Leeuwen, M. Rem (Eds.), PARLE '91. Parallel Architectures and Languages Europe, Volume I. Proceedings, 1991. XV, 423 pages. 1991.

Vol. 506: E. H. L. Aarts, J. van Leeuwen, M. Rem (Eds.), PARLE '91. Parallel Architectures and Languages Europe, Volume II. Proceedings, 1991. XV, 489 pages. 1991.

Vol. 507: N. A. Sherwani, E. de Doncker, J. A. Kapenga (Eds.), Computing in the 90's. Proceedings, 1989. XIII, 441 pages. 1991.

Vol. 508: S. Sakata (Ed.), Applied Algebra, Algebraic Algorithms and Error-Correcting Codes. Proceedings, 1990. IX, 390 pages. 1991.

Vol. 509: A. Endres, H. Weber (Eds.), Software Development Environments and CASE Technology. Proceedings, 1991. VIII, 286 pages. 1991.

Vol. 510: J. Leach Albert, B. Monien, M. Rodríguez (Eds.), Automata, Languages and Programming. Proceedings, 1991. XII, 763 pages. 1991.

Vol. 511: A. C. F. Colchester, D.J. Hawkes (Eds.), Information Processing in Medical Imaging. Proceedings, 1991. XI, 512 pages. 1991.

Vol. 512: P. America (Ed.), ECOOP '91. European Conference on Object-Oriented Programming. Proceedings, 1991. X, 396 pages. 1991.

Vol. 513: N. M. Mattos, An Approach to Knowledge Base Management. IX, 247 pages. 1991. (Subseries LNAI).

Vol. 514: G. Cohen, P. Charpin (Eds.), EUROCODE '90. Proceedings, 1990. XI, 392 pages. 1991.

Vol. 515: J. P. Martins, M. Reinfrank (Eds.), Truth Maintenance Systems. Proceedings, 1990. VII, 177 pages. 1991. (Subseries LNAI).

Vol. 516: S. Kaplan, M. Okada (Eds.), Conditional and Typed Rewriting Systems. Proceedings, 1990. IX, 461 pages. 1991.

Vol. 517: K. Nökel, Temporally Distributed Symptoms in Technical Diagnosis. IX, 164 pages. 1991. (Subseries LNAI).

Vol. 518: J. G. Williams, Instantiation Theory. VIII, 133 pages. 1991. (Subseries LNAI).